Environmental Design

Written and edited by a team of special... ...rdham LLP, one of the UK's leading environmental and building serv... ...neering consultancies, *Environmental Design* is the result of their extensive experience in designing environmentally friendly buildings. The principles of their approach, which they have taught in numerous schools of architecture and engineering, are clearly presented here. This book is essential reading for architects, engineers, planners and students of these disciplines and for all those who are concerned with our built environment.

The book starts with some basic scientific principles and issues of air quality, noise and climate change and then moves on to site planning, energy use, materials and building form. Natural ventilation systems, high-efficiency mechanical equipment and alternative energy sources are covered. State-of-the-art buildings of exceptional quality are incorporated throughout the text and illustrate the authors' belief that environmentally responsible architecture can be visually exciting. This extensively updated third edition concludes with a selection of detailed case studies of award-winning projects – including, for the first time, Beaufort Court, King's Langley and Heelis, central office for the National Trust, Swindon.

Randall Thomas has been active in practice and teaching for over thirty years. He is visiting Professor of Sustainable and Environmental Design at Kingston University, UK, and has been a Partner of Max Fordham LLP since 1987.

Environmental Design

An introduction for architects and engineers

Third edition

Edited by

Randall Thomas

Max Fordham LLP

Taylor & Francis
Taylor & Francis Group

LONDON AND NEW YORK

First published 1996 by E & FN Spon
This edition published by Taylor and Francis
2 Park Square, Milton Park, Abingdon, Oxon OX14 4RN

Simultaneously published in the USA and Canada
by Taylor & Francis Inc
270 Madison Ave, New York, NY 10016

Taylor & Francis is an imprint of the Taylor & Francis Group

Typeset in Ehrhardt by RefineCatch Limited, Bungay, Suffolk
Printed and bound in Great Britain by Bell & Bain Ltd, Glasgow

British Library Cataloguing in Publication Data
A catalogue record for this book is available from the British Library

Library of Congress Cataloging in Publication Data
Environmental design : an introduction for architects and engineers /
edited by Randall Thomas.—3rd ed.
    p. cm.
  Includes bibliographical references and index.
    ISBN 0–415–36330–0 (alk. paper)—ISBN 0–415–36334–9 (pbk. :
alk, paper)—ISBN 0–203–01366–2 (e-book)
    1. Architecture—Environmental aspects.   2. Environmental engineering.
3. City planning.   4. Engineering design.   I. Thomas, Randall.
NA2542.35.E575   2005
720′.47—dc22   2005049928

ISBN10: 0–415–36333–0 (hbk)
ISBN10: 0–415–36334–9 (pbk)

ISBN13: 9–78–0–415–36333–0 (hbk)
ISBN13: 9–78–0–415–36334–9 (pbk)

CONTENTS

Contributors   vii

Foreword by Edward Cullinan   ix

Preface by Max Fordham   xi

Apologies and acknowledgments   xiii

Units and abbreviations   xv

Part One   1

1 Strategies   3
   MAX FORDHAM

2 Comfort, health and environmental physics   7
   BILL WATTS

3 Buildings and energy balances   29
   BILL WATTS AND RANDALL THOMAS

4 Building planning and design   37
   RANDALL THOMAS

5 Site planning   54
   RANDALL THOMAS

6 Materials and construction   67
   RANDALL THOMAS

7 Energy sources   81
   RAMIRO GODOY

8 Lighting   96
   RANDALL THOMAS

9 Engineering thermal comfort   110
   RANDALL THOMAS

10 Water, waste disposal and appliances   145
   MIKE ENTWISLE, RANDALL THOMAS AND ADAM RITCHIE

   Summary   153
   RANDALL THOMAS

Part Two   155

11 RMC International Headquarters   157
   BART STEVENS

12 Queens Building, De Montfort University   165
   EDITH BLENNERHASSETT

13 The Charles Cryer Studio Theatre   183
   COLIN HAMILTON

14   The Environmental Building, Building Research Establishment   190
      RANDALL THOMAS AND BART STEVENS

15   The Millennium Centre, Dagenham   211
      RANDALL THOMAS

16   The Bedales Theatre   217
      RANDALL THOMAS

17   Beaufort Court   222
      TAMSIN TWEDDELL

18   Heelis, Central Office for the National Trust   231
      GUY NEVILL

Appendices   239

A   Environmental data and the psychrometric chart   241
     RANDALL THOMAS

B   Calculation Procedures   249
     RANDALL THOMAS

C   Acoustics   253
     RANDALL THOMAS AND ANTHONY CHILTON

D   Photovoltaics   255
     RANDALL THOMAS AND LORNA MARKHAM

Illustration acknowledgements   262

Index   265

CONTRIBUTORS

Edith E. Blennerhassett, BE, MIEI, Former Partner, Max Fordham Associates

Anthony Chilton, MA, D Phil

Mike Entwisle, MA, PhD, Former Partner, Max Fordham & Partners

Max Fordham, OBE, MA, FEng, FCIBSE, MConsE, Hon FRIBA, Visiting Professor in Building Environmental Design, University of Bath, Partner, Max Fordham LLP

Ramiro Godoy, MSc, DipIng, CEng, MCIBSE, Partner, Max Fordham LLP

Colin Hamilton, BSc, Partner, Max Fordham LLP

Lorna Markham, MEng

Guy Nevill, MEng, Partner, Max Fordham LLP

Adam Ritchie, BSc, CPhys, MInstP, Partner, Max Fordham LLP

Bart Stevens, BSc, CEng, FCIBSE, FEI, Partner, Max Fordham LLP

Randall Thomas, PhD, EurIng, CEng, FCIBSE, FConsE, MASHRAE, Visiting Professor in Architectural Science, Kingston University, Partner, Max Fordham LLP

Tamsin Tweddell, MA, MSc, CPhys, MInstP, Partner, Max Fordham LLP

Bill Watts, BA, MSc, CEng, FCIBSE, Partner, Max Fordham LLP

FOREWORD

This book was first published ten years ago and I provided a foreword then. In the intervening years, during which environmental design has more and more been seen as sustainable design (for good earth preserving reasons), two things have happened.

Most buildings have more or less reluctantly taken on board ever increasing legal requirements for greater insulation (heat and sound), reduced glazing, improved orientation, the use of less wasteful materials and more natural ventilation; while all these things are not seen as having any implications for the architectural language that is used. This is the eternal no change school of behaviour.

But a minority have struggled to find a new or revised architectural expression to allow for sustainable environmental conditions. Architects, engineers and designers have searched for a new aesthetic for such things as solar shading, load bearing masonry, green wood building, natural (chimney effect) ventilation, wind and photo-voltaic power generation, passive solar gain, and so on.

For it is hard to see how such considerations might be incorporated without a change in the way that we think about architecture. It is hard to see how to make that change without thinking about the composition of architecture in a more expressive way than is common today. If we are to make this move towards responsive expressionism we will need all the help we can get from clear thinking, cool headed environmental designers. It is here that this book *Environmental Design* has a most important part to play. It is clear, logical, well illustrated and good to read; and it has the great quality of all profound work – it is easy to understand. Now let us use it to help us with our architecture.

Edward Cullinan
January 2005

PREFACE

We here present another edition of this book.

I have a feeling of urgency. The economies of the world are growing at an unprecedented rate and delivering the benefits of industrialisation in a way that is necessary to lead to a stable global population during this century. This growth will be a disaster if it carries an equivalent emission of carbon dioxide. During this century architecture must reduce the demand for energy while maintaining the benefits of health and comfort.

This book provides the basic information to implement the new needs.

'Good luck!' is not enough. Take up the challenge.

Max Fordham

APOLOGIES AND ACKNOWLEDGEMENTS

We owe an apology to Matisse for the woman in robes in Figure 2.1 and to Rembrandt for the elephant of Figure 4.9. A number of building outlines are loosely based on projects by Le Corbusier and Philip Johnson whom the editor admires.

We would like to thank again all the individuals and organizations who supported the first two editions of this book. The current edition has only been made possible by Juliet Baptiste who applied both her editorial and political talents to a difficult task, Jules Mohm and Caroline Mallinder of Taylor and Francis. Thanks also to Tony Leitch who has supplied us with further illustrations for our updated text.

Many people have helped with updating the work. Our thanks to Anna Georgiou, Kitty Lux, Emilia Melville, Alasdair Reid, Scott Rushford and Ian Smith.

Special thanks to Feilden Clegg Bradley Architects who once again have provided us with a magnificent building for the cover.

The students we have taught in schools of architecture and engineering throughout the UK and France have helped refine our ideas.

We cannot thank our clients, and the architects with whom we work, too much. Without them the book would never have been published.

Randall Thomas

UNITS AND ABBREVIATIONS

1. Physics and units

The SI (Système Internationale) unit of force is the newton and the unit of work (force times distance) is the newton-metre, also defined as a joule. Work and energy have the same units. Power is the amount of energy expended (or work done) per unit time – one joule per second is a watt. If a 100 watt bulb is left on for one hour the energy consumption is 100 watt-hours. This in turn can be expressed as 360 000 J (since one watt-hour equals 3600 J). Heat is a form of energy and has the same units.

Pressure is the force acting per unit area. One pascal (Pa) is a force of one newton (N) per square mctre (m^2).

The unit of thermodynamic temperature in the SI system is the kelvin (K). For this reason derived units such as thermal conductivity are expressed as watts per metre kelvin (W/m K). However, the Celsius (°C) temperature scale is also in common use (the Celsius scale is also known as the centigrade scale). Absolute temperature in degrees kelvin is found by adding 273 to degrees Celsius. Thus,

$$30 \ ^\circ C + 273 = 303 \ K$$

The light emitted by a source (or received by a surface) is the luminous flux. The SI unit of luminous flux is the lumen. Illuminance is the luminous flux incident per unit area. One lumen per square metre is one lux.

2. Conversion factors

Length
1 micron $= 1 \times 10^{-6}$ m
1 m $\quad = 3.281$ ft

Area
1 $m^2 = 10.76$ ft^2

Volume
1 $m^3 = 35.31$ ft^3

Mass
1 kg $= 2.205$ lb

Force
1 N $\cong 0.1$ kg (force)
1 N $\cong 0.22$ lb (force)

Pressure
1 Pa $\quad = 0.004$ in H_2O
1 kPa $\quad = 0.145$ psi (lb/in^2)
1 bar $\quad = 100 \ 000$ Pa

Energy, work, heat
1 kJ	= 0.948 Btu
1 MJ	= 0.278 kWh
1 GJ	= 278 kWh
1 therm	= 105.5 MJ

Power
1 kW = 1.341 hp

Thermal conductivity
1   W/m   K = 0.578 Btu/(ft h °F)

Heat transfer coefficient
1 W/m^2 K = 0.176 Btu/(ft^2 h °F)

Temperature
K	= 273 + °C
°C	= (5/9)(°F − 32)

Temperature intervals
1 °C	= 1.8 °F
1 °C	= 1 K

3. Abbreviations

AC	= alternating current
AHU	= air-handling unit
BMS	= building management system
BRE	= Building Research Establishment
ca	= circa; approximately
CIBSE	= Chartered Institute of Building Services Engineers
db	= dry bulb
HWS	= hot water service
M	= million; mega (i.e. 1×10^6)
nm	= nanometres
r.h.	= relative humidity
rev/min	= revolutions per minute
SI	= Système Internationale
wb	= wet bulb
yr	= year

4. Further reading

Duncan, T. (1994) *Advanced Physics: Materials and Mechanics* (4th edition), John Murray, London.

Part One

Strategies

1.1 Introduction

Environmental design is not new. The cold environment of 350 000 years ago led our European ancestors to build shelters under limestone cliffs (Figure 1.1). More recently, English cob cottages (Figure 1.2) and Doha homes (Figure 1.3), both built of earth, demonstrate vernacular responses to light and heat.

The cob cottage evolved to provide sufficient light (say about 100 lux) under overcast skies and to limit heat loss in the winter. The Doha home evolved to provide about the same light level in bright sunlight while protecting the interior from extreme heat. It is no coincidence that the two buildings developed to give the same light level.

The rise of science in the Renaissance led to the Industrial Revolution which has enabled environmental engineers to produce reasonably comfortable conditions in almost any building in almost any climate. Some of the most visually powerful architecture of our era has taken technology and pushed it to the limits of its capabilities. The engineering systems associated with this architecture, however, have required high-grade energy to deal with the environmental problems resulting from the building design. What we need to do now is to reduce a building's reliance on fossil fuel-derived high-grade energy yet still provide comfort inside for the occupants.

What is important in achieving this? Solar energy drives the processes we live by – photosynthesis, the carbon cycle, weather, the water cycle. It is the source of our fossil fuels and it sustains the average world temperature at about 15 °C. The light from the Sun can replace the high-grade energy used in electric lighting. One watt of natural

1.1 Reconstruction of an Acheulean hut in France.[1]

1.2 English cob cottage.[2]

light more than replaces three watts of primary energy used by a fluorescent light and even more if replacing wasteful tungsten light bulbs. Wind is invaluable in providing fresh air and in helping to lower summer-time temperatures.

Our activities as a species are likely to overload our habitat. We plunder fuel reserves and convert them to carbon dioxide as we generate the energy for our immediate use. Our other chemical activities can pollute the environment, as we have seen with the depletion of ozone in the stratosphere. We must change to designing our buildings to reduce our impact on the global environment.

In general, buildings are like many animals. In cold weather a source of energy is needed to keep them up to temperature and a strategy is required to prevent the heat inside from escaping. In bears, this is accomplished by a good coat of fur – in buildings it may be by insulation. The heat loss associated with ventilation also requires regulation and much recent work deals with better seals and control techniques.

In hot weather, when the external temperature is high, too much heat may enter the space. If this heat can be absorbed by the fabric of the building, the peak air temperature during the day will be less. If night-time ventilation is possible, the heat absorbed by the fabric of the building can be lost at night when the temperatures are lower. But if the buildings are lightweight and sealed, they are likely to overheat and a need for air-conditioning will result.

What do we need to do now?

Recently, the most significant shift in thinking is to consider the building as a whole. From this perspective we should examine how the site, form, materials and structure can be used to reduce energy consumption but maintain comfort. The importance of

1.3 Doha home.[3]

daylight has also become more clear. Natural light is our most important and reward-ing use of solar energy. Building design should aim to provide enough light whenever the Sun is above the horizon. Of course, the dangers of glare and overheating must be avoided; therefore façade design and ventilation are key elements to a successful strategy. As part of this we need to develop economic triple-glazed windows with automatically (or easily) operated blinds to control solar radiation during the day. These blinds should be thermally insulated (or they might even be separate shutters) to reduce heat loss at night. With regard to ventilation, air quality and noise will be major design factors which, in critical situations, may lead us back to fans and simple mechanical systems.

The following chapters of this book deal with these and other issues that will help us manage the transition to comfortable, healthy, well-designed, energy efficient buildings. It is based on the leading edge of current practice but it will be evident to the reader that more buildings, more monitoring and more data are needed. The book gives pointers but it is no substitute for thought – thinking about buildings has a long way to go.

Notes and references

1. Musée Régional de Préhistoire, Orgnac l'Aven, France.
2. Clifton-Taylor, A. (1972) *The Pattern of English Building*, Faber & Faber, London.
3. Illustration and photograph by Max Fordham.

Comfort, health and environmental physics

2.1 Introduction

This chapter discusses human comfort. Part of the background for the topic is an introduction to basic scientific principles of heat transfer, the electromagnetic spectrum, light and sound and their relationship to building design. It concludes with an overview of air quality, both outdoor and indoor, and moisture issues in buildings.

2.2 Comfort and control

As an example, let us consider our own bodies, which are controlled to maintain a core temperature close to 37 °C. We function best at this temperature and variations on either side are detrimental. This temperature has evolved over a very long period under the influence of many variables.

A major factor is the need to get rid of the heat we generate as a by-product of our metabolic systems. The heat we produce varies from about 100 W at rest to about 1000 W when physically very active. A seated adult male indoors in normal conditions produces about 115 W – about 90 W of which is sensible heat and the remaining 25 W is latent heat. Sensible heat is that which we can 'sense' or feel; it is detectable through changes in temperature. Latent heat is the heat taken up or released at a fixed temperature during a change of phase, e.g. from a liquid to a gas.

Heat loss from the body occurs in several ways. Sensible heat loss from the skin or outer clothing surface occurs by convection and radiation, and there is a sensible heat loss during respiration. Latent heat loss occurs through the evaporation of sweat, and the evaporation of moisture during respiration. The rate of heat loss from the body will depend on the air temperature, the mean radiant temperature of the surroundings (for example, in a room this is the mean of the temperatures of the walls, glazing, ceiling and floor), the air speed and the clothing worn. In temperate climates, the atmospheric water vapour pressure (i.e. the pressure exerted by the water vapour component of the air) has a slight effect on heat flow from the body;[1] in hot, humid situations the effects can be much more significant.

The naked body, if shaded from the Sun, can be quite comfortable at around 28–30 °C and at moderate relative humidities (see Appendix A for a definition) of, say, 50%. As the ambient temperature rises the body's response is (a) to direct more blood to the surface, which increases the skin temperature and heat loss, and (b) to sweat to lose heat through evaporation. We begin to feel uncomfortable when these responses become significant.

When the ambient temperature drops, the body will limit heat loss by reducing blood flow to the surface, which reduces the skin's temperature, and by not sweating. Goose pimples are caused by tiny muscles lifting hairs on the skin, which will decrease the air flow and heat loss across the surface. This mechanism was more effective in our more hairy ancestors, but we have more than compensated through the use of clothing. Indeed the ability to clothe ourselves is considered a reason for our success over other hominids. With air temperatures between, say, 20 and 26 °C we limit our heat loss with clothing but generally feel comfortable. The amount of clothing we require increases as the temperature decreases and the general atmosphere becomes less comfortable. If the heat loss through our clothing becomes too great, we generate more heat specifically for temperature control by shivering.

Two points here are relevant to building design. Firstly, to make buildings comfortable, they should be kept within a suitable temperature range which is not as wide as that in an uncontrolled external environment. Secondly, our bodies are capable of maintaining a very stable core temperature with a fairly constant metabolic heat output over a wide range of external temperatures. This is done, with little or no additional energy expenditure, by a combination of control processes including sweating, altering the blood flow (and therefore the heat loss to the skin) and changing clothes to suit conditions.

Modern buildings have achieved the first objective of maintaining fairly constant internal conditions to comfort standards with the use of significant amounts of energy to provide heating or cooling to compensate for the changing external environment. The amount of energy used could be reduced significantly if buildings adopted the principles of animal physiological control. To do this one first needs an understanding of human comfort and how energy is expended to provide it.

Comfort is a subjective matter and will vary with individuals. It involves a large number of variables, some of which are physical with a physiological basis for understanding. Classically, for thermal comfort they include:

– air temperature and temperature gradients
– radiant temperature
– air movement
– ambient water vapour pressure
– amount of clothing worn by the occupants
– occupants' level of activity.

Other factors influencing general comfort are light levels, the amount of noise and the presence of odours. Individuals are also affected by such psychological factors as having a pleasant view, having some control of their environment and having interesting work. For some variables it is possible to define acceptable ranges but the optimal value for these will depend on how they interact with each other, e.g. temperature and air speed, and personal preference.

2.3 Thermal comfort and heat transfer

To be thermally comfortable one must not feel too hot or too cold, or have any part of the body too hot or too cold. The physiological basis for this is that the amount of heat being produced by the body is in balance with the heat loss, comfortably within the body's control mechanism. Furthermore, there should be no parts of the body having to operate outside the comfortable limits of the control systems (such as a cold neck from a draught or a hot face from an open fire).

There are several mechanisms which transfer heat and therefore affect this balance. Heat always flows from hot bodies to cold ones. Individuals may be gaining or losing heat depending on the relative temperatures of their bodies and their surroundings. They may indeed be gaining heat through one mechanism and simultaneously losing it through another.

As an illustration, on a very hot day in the desert the air and sand are both likely to be hotter than a person's body. The body is therefore heated by the sand on which one stands, by the air and by the Sun. Depending on the exact air and body temperatures, the only cooling mechanism may be by evaporation through respiration and sweating. Figure 2.1 illustrates the four heat transfer processes of conduction, convection, radiation and evaporation/condensation. These basic physical processes apply both to humans and buildings, and we shall examine them in turn.

Conductive heat transfer and thermal mass

In this process heat travels through matter by one hot vibrating molecule shaking the cooler ones adjacent to it, thereby making them hotter and passing the heat along (or, more technically, by the transfer of kinetic energy between particles). Metals tend to be good conductors and so have high thermal conductivities. (The thermal conductivity of a material is the amount of heat transfer per unit of thickness for a given temperature difference.) Organic materials such as wood and plastic tend to be poor conductors. Aerated materials, which have solid conduction paths broken by air or gas gaps (foam, glass fibre quilt or feathers) are very poor conductors but are good insulators as they have low thermal conductivities. Table 2.1 shows a range of thermal conductivities and other properties of some materials.

Crudely, if an object is cold to touch it is probably a good conductor as it conducts the heat away from your hand. If it is warm it is a good insulator as the heat is not drawn from your hand. This, of course, assumes that the object is colder than your skin temperature. Hotter conducting materials will feel hotter than insulating ones as more heat will be conducted to your hand.

One of the pitfalls of this test illustrates another property of materials: the capacity to store heat. At room temperature aluminium foil may feel cold for a short period, but then warm. This is because the small mass involved heats up quickly and not because it is a bad conductor. A thick aluminium saucepan would feel cold as there is adequate mass to absorb a reasonable quantity of heat. Similarly, one may risk holding a loose piece of foil straight from the oven with bare hands, but not a solid baking dish.

The amount of heat a material can store, or its thermal mass, is the product of multiplying its mass, its specific heat capacity and the increase in temperature. The specific heat capacity (Table 2.1) is the amount of heat that a material will store per

[a] The loose-fitting robes touch the skin at the shoulders only. Depending on the exact air and body temperatures an upward draught of air can help keep the wearer cool by increasing the rate at which sweat evaporates.[2] This thermally induced upward draught is known as stack-effect ventilation and is also common in buildings (Chapter 9).
[b] Approximate temperatures.

2.1(a) Heat transfer mechanisms: the Bedouin by day.

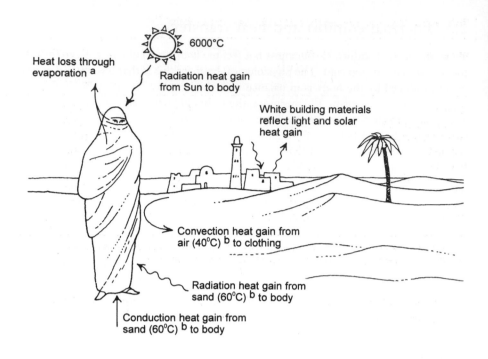

Heat loss through evaporation [a]

6000°C

Radiation heat gain from Sun to body

White building materials reflect light and solar heat gain

Convection heat gain from air (40°C) [b] to clothing

Radiation heat gain from sand (60°C) [b] to body

Conduction heat gain from sand (60°C) [b] to body

[a] Evaporative heat losses occur from the skin and respiration. At night clothing is used, especially for its insulating effect.
[b] The night sky has an effective temperature of −45 °C for radiation from the Earth.[3]

2.1(b) Heat transfer mechanisms: the Bedouin at night.

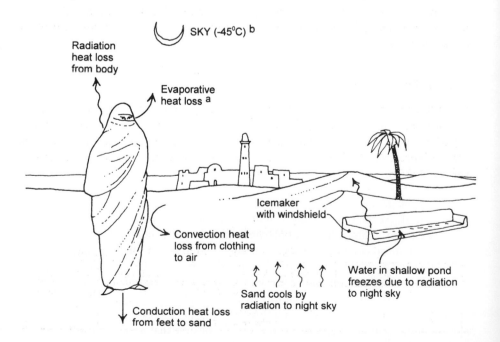

SKY (−45°C) [b]

Radiation heat loss from body

Evaporative heat loss [a]

Convection heat loss from clothing to air

Icemaker with windshield

Water in shallow pond freezes due to radiation to night sky

Sand cools by radiation to night sky

Conduction heat loss from feet to sand

Table 2.1 Typical properties of selected materials[4,5]

Material	Density (kg/m³)	Thermal conductivity (W/m K)	Specific heat capacity (J/kg K)
Bricks	1700	0.73[a]	800
Concrete, dense	2000	1.13	1000
Glass fibre quilt	25	0.035	1000
Asphalt	1700	0.50	1000
Aluminium	2700	214	920
Water (20 °C)	1000	0.60	4187
Sand (dry)	1500	0.30	800
Steel	7800	45	480
Wood	500–700[b]	0.12–0.23[b]	1200–3400[b]

[a] Mean of internal and external brick types. Consult manufacturers' data for precise values.
[b] Values vary depending on wood type, temperature and water content.

unit of mass and per unit of temperature change. Note the comparatively high specific heat capacity of water.

Convective heat transfer

Convective heat transfer is the process by which heat is transferred by movement of a heated fluid such as air or water. If we consider a hot surface and a cold fluid, the fluid in immediate contact with the surface is heated by conduction. It thus becomes less dense and rises, resulting in what are known as natural convection circulation currents. Convection that results from processes other than the variation of density with temperature is known as forced convection and includes the movement of air caused by fans.

Radiant heat transfer

In conduction and convection, heat transfer takes place through matter. For radiant heat transfer, there is a change in energy form, and bodies exchange heat with surrounding surfaces by electromagnetic radiation such as infrared radiation and light (Figure 2.2). Surfaces emit radiated heat to, and absorb it from, surfaces that surround them.

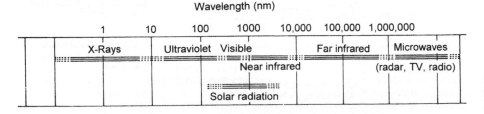

2.2 The electromagnetic spectrum. (See also Figure 2.3.)

The amount of heat emitted from a surface depends on its emittance and its temperature; as the temperature increases the heat emitted increases (it is, in fact, proportional to the fourth power of the absolute temperature). The emittance itself is a function of the material, the condition of its surface, wavelength and the temperature. The amount of radiation absorbed by a surface is its absorptance. The absorptance depends on the nature of the surface, the spectral distribution of the incident radiation and its directional distribution. (In contrast with emittance, the temperature of a surface has only a very small effect on its absorptance.) Table 2.2 gives some data on emittances and absorptances.

As can be seen, most surfaces commonly used in buildings (paint, dull metal, glass, etc.) have high emittances (about 0.8–1). Black non-metallic surfaces can be seen to have both high emittances and absorptances. Shiny metallic surfaces have both low emittances and absorptances.

When radiation strikes a surface it is either absorbed, reflected or transmitted (i.e. it passes through the material struck) with the relative proportions depending on the characteristics of the surface and the wavelength of the incoming radiation. Absorbed radiation will, of course, cause a material to heat up.

The varying nature of materials can be exploited – for example, heat loss through windows can be reduced by using a low emissivity coating on the glass; solar collectors use films with high absorptance and low emissivity to maximize solar gain.

Table 2.2 Emittances and absorptances of selected materials[6]

Item	Emittance (at 10–40 °C)	Absorptance (for solar radiation)
1. Black non-metallic surfaces such as asphalt, carbon, slate, paint	0.90–0.98	0.85–0.98
2. Red brick and tile, concrete and stone, rusty steel and iron, dark paints (red, brown, green, etc.)	0.85–0.95	0.65–0.80
3. Yellow and buff brick and stone, firebrick, fireclay	0.85–0.95	0.50–0.70
4. White or light cream brick, tile, paint or paper, plaster, whitewash	0.85–0.95	0.30–0.50
5. Bright aluminium paint; gilt or bronze paint	0.40–0.60	0.30–0.50
6. Polished brass, copper, monel metal	0.02–0.05	0.30–0.50
7. Low emissivity glass	0.04[a]	0.11–0.48[b]
8. Solar thermal collector's coating	0.03–15[c] 0.2[d]	0.92–0.98[c] 0.9[d]

[a] See reference 7.
[b] See reference 8.
[c] Based on selective absorber layers such as Black Chrome on Nickel or Aluminium-Nitrogen on Aluminium. Note production tolerances of individual products may vary considerably. See reference 9.
[d] Based on non-selective absorber layers such as solar paint. See reference 10.

For people, the mean radiant temperature of their surroundings is important for comfort, as is the variation or uniformity of radiant temperature – imbalances that make one hot and cold on different sides can be disagreeable. Radiant heat transfer can be felt most obviously, for example, by standing outside facing a bonfire on a clear night. One's face will be hot from the radiation of the fire and one's back (if lightly clad) will feel colder, partly because of convective heat transfer to the cool night air and partly from body heat being radiated to the surrounding night sky (Figure 2.1(b)).

Evaporative heat transfer

Molecules in a vapour state contain much more energy than the same molecules in a liquid state. Thus energy must be added to turn a liquid into a gas. The amount of heat required to change liquid water into a vapour is the latent heat of evaporation. This heat is removed from the liquid – which is thus cooled – and transferred to the vapour. Evaporation cools a wet surface.

In condensation the process is reversed and the latent heat of evaporation is transferred from the vapour to the surface.

The amount of energy transferred in evaporation and condensation is considerable compared to that required to heat or cool a liquid or gas; for example, to vaporize 1 kg of water at boiling point it takes about 500 times as much heat as is required to increase the temperature of 1 kg of water by 1 °C. Steam at 100 °C is more likely to burn one's skin than dry air from a hot oven at over 200 °C.

The maximum amount of water vapour that can be held in a fixed mass of air is related to the temperature of the air (Appendix A). At 20 °C it can carry up to 15 g of water per kg of air; at 0 °C it can hold only 4 g of water.

The direction of vapour flow to or from a wet surface is dependent on the quantity of water in the air and the temperature of the surface. The air immediately above a wet surface is assumed to be at the same temperature as that surface, and 100% saturated. This will define a water vapour content in grams of water per kilogram of dry air. For that surface to lose heat by evaporation, the surrounding air must have a lower amount of water vapour per kilogram of dry air. In very hot and humid conditions the water content of the air is high and so the surface temperature must be comparatively higher to lose heat. The rate of evaporation is determined by the difference in water vapour content and the air speed across the surface.

Thermal comfort levels

All the processes described above contribute to our thermal balance, which is the sum of the effects of the heat exchanged by the body with its environment. We feel comfortable if we can maintain our thermal balance without much effort, and uncomfortable in our environment if we have to shiver to generate heat or sweat profusely to lose it.

It is all about the balance between activity levels and the heat we produce, how well insulated (clothed) we are, and the thermal environment we are in. For thermal environment read air temperature, radiant temperature, humidity and air movement. This family of relationships has been well researched. However, we still tend to judge the thermal environment as simply the temperature according to a thermometer on the wall. In many cases this is sufficient, but an extreme in any one of the other variables will cause the situation to change.

In any one room there may be a wide range of air temperatures caused by hot air at high level or cool air at a lower level leading to a hot head and cold feet. High air movement and the turbulence of the air can exacerbate the discomfort of cold drafts or provide relief from over heating in a hot environment. Cold radiation from a window will be a problem when it is cold, as will sitting in the sun when it is hot. However a chilled ceiling will cool a space during the summer and powerful radiation from the sun can make a cold day feel warmer. Finally, drier air will feel cooler than more humid air due to the greater ability of the body to lose heat by sweating.

One can easily control all of these variables by using engineering systems that consume energy. However, we are now designing buildings to use substantially less energy and as a result the environmental conditions will be less tightly controlled. One is now relying on intelligent design to manage the variety of factors in order to keep people comfortable.

Most people will feel comfortable at 21°C doing sedentary work dressed in a suit or a sweater etc. At 26–27°C, we need to wear lighter clothes. Clearly dress codes need to reflect this change in temperature.

In the UK, summer daytime temperatures are usually in the range of 21–25°C, however on occasions they can increase to 32–34°C for short periods. Without mechanical cooling it is difficult to design a building that will stay below 27°C when it is 32°C outside. This is recognised by design guides that allow the temperatures to exceed a fixed temperature for a limited period of time within a given model year. The client or occupier must be made aware, understand and accept these choices. For schools the UK Department for Education and Skills (DfES) has a similar calculation based on a school year, allowing the temperature to exceed 28°C for 80 hours during the school term.[11]

Overheating can be alleviated with air movement provided by an open window, however this can conflict with external noise and cause problems with internal papers rustling. It may be possible to deal with this by design but individuals prefer the ability to manage their window and choose between noise and heat levels.

2.4 The electromagnetic spectrum

Electromagnetic radiation is energy in the form of waves generated by oscillating magnetic and electrical fields. This radiation covers a spectrum of wavelengths, as shown in Figure 2.2.

The spectrum has no definite upper or lower limit and regions overlap; the visible region is roughly from 400 to 760 nm.

The wavelength multiplied by the frequency equals the speed of propagation, which is the speed of light. The energy of the radiation is proportional to its frequency. Intensity of radiation decreases with the square of the distance from a point source; thus, if the distance from the source is doubled, the intensity falls to one-quarter.

All matter warmer than absolute zero (0 K or −273 °C) produces a spectrum of radiation which varies with its temperature. A blackbody is given this term because it absorbs all the energy incident on it; these perfect absorbers are also perfect radiators, or emitters, of energy. (Most bodies are in fact less perfect and are treated as grey bodies.) The hotter the body, the more total energy is radiated and the higher the energy and frequency of the radiation; correspondingly, the wavelength of the emitted energy is lower (Figure 2.3). The Sun, whose spectrum is essentially that of a blackbody at 6000 °C, produces a broad range of radiation from ultraviolet through visible light to infrared. A filament light bulb produces most of its energy as invisible infrared radiation with some visible light. This is why a filament lamp is comparatively inefficient compared to a fluorescent one, which is designed to produce most of its radiation in the visible spectrum.

Figure 2.4 shows the spectral distribution, i.e. the amount of radiation at various wavelengths, of solar radiation at the Earth's surface. The atmosphere filters out most of the Sun's ultraviolet and much of the infrared radiation; the ultraviolet being particularly removed by the ozone layer in the upper atmosphere. At the Earth's

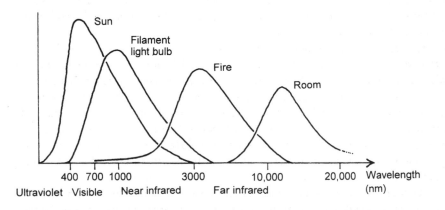

2.3 Spectra of radiation from the various bodies.[12] (*N.B.* The vertical axis is not to scale and does not give relative intensity.)

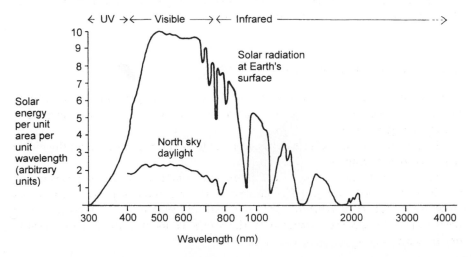

The approximate spectral composition and intensity of north sky daylight of 5700 K is also plotted for comparison (see Figure 8.7).

2.4 Spectral distribution of solar radiation at the Earth's surface.[13]

surface, approximately 10% of the Sun's radiated energy is in the ultraviolet range of 290–400 nm, 40% in the visible range of 400–760 nm and 50% in the infrared range of 760–2200 nm.[14] (These figures vary somewhat with how the wavelength band radiation limits are defined and according to which reference one consults.)

Life on Earth has evolved to make use of incident solar radiation – human vision and photosynthesis being two examples of this. The depletion of the ozone layer has become an area of major concern. We and other forms of life have not developed biological mechanisms to protect us from large amounts of high-energy ultraviolet (UV) radiation as most of this is filtered out by the ozone layer. The likely consequences of continued loss of ozone in the atmosphere include a higher incidence of skin cancer and eye cataracts, and damage to land and marine vegetation by UV radiation. Chemicals, mainly chlorofluorocarbons (CFCs) and hydrochlorofluoro-carbons (HCFCs), which are depleting the ozone layer, are discussed in Chapter 6; CFCs and HCFCs also contribute to global warming as discussed below and as shown in Figure 2.5. Their relative contribution to global warming can be defined by their Global Warming Potential (GWP) which is defined as 'the radiative warming effect of 1 kg of each gas with respect to carbon dioxide for a specified period of time subsequent to the emission'.[15]

Water vapour, carbon dioxide and ozone in the atmosphere absorb infrared radiation from the Sun and from the surface of the Earth. This insulates the planet and keeps it warm – unlike the surface of the Moon which swings greatly in temperature, varying from 100 °C on the sunlit surface to −150 °C at night. The insulating effect of water vapour in clouds can be seen when comparing a clear night to a cloudy one. On a cloudy night the ground cools down comparatively slowly as it is radiating to a thick blanket of water vapour which is absorbing the heat and radiating much of it back to Earth. On a clear night the sky has less water vapour, and thus much more of the infrared radiation from the ground escapes into space.

The Earth will, on average, lose the same amount of heat to the Universe as it gains from the Sun. If the atmosphere becomes a better insulator of infrared radiation, the Earth's surface will become warmer in order to lose the same amount of heat coming from the Sun. There is great concern that higher 'greenhouse' gas levels and, in

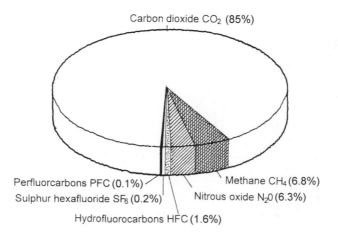

Carbon dioxide CO_2 (85%)

2.5 Total emissions of UK greenhouse gases weighted by Global Warming Potential 2002.[16]

Perfluorcarbons PFC (0.1%)

Sulphur hexafluoride SF_6 (0.2%)

Methane CH_4 (6.8%)

Nitrous oxide N_2O (6.3%)

Hydrofluorocarbons HFC (1.6%)

particular, carbon dioxide from burning fossil fuels (Figure 2.5) are subtly increasing the insulation of the atmosphere, thus causing a rise in the world's average temperature. It is feared that a small increase in temperature will melt sufficient amounts of water currently frozen in polar ice-caps to raise the ocean's water level to flood low-lying land all over the globe.

2.5 Light

As mentioned above, light is the visible portion of the electromagnetic spectrum. Wavelengths are associated with colours, as can be seen in Figure 2.6, which also shows that the eye is most sensitive to green light at about 550 nm. White light is a mixture of various wavelengths.

Light levels outside vary enormously from 100 000 lux in bright sunlight to 0.2 lux in bright moonlight and 0.02 lux in starlight. Our eyes will register information over the range of brightest sunlight down to about 0.005 lux.

While this is our full range, the amount and content of information we can register drops off at low light levels. A young person with good eyes can probably thread a needle in 100 lux, read a theatre programme in 10 lux and distinguish large objects in 0.005 lux. It would be a strain if one was always to do these tasks at these light levels and older people would find them difficult to do at any time. For this reason the recommended light levels associated with various tasks tend to be 5 to 10 times greater than the absolute minima (Chapter 8).

Our eyes can deal with the range of lighting levels by adapting to different average ambient light levels. They cannot, however, register information across the entire range at the same time as they take time to adapt when going from a light space to a dark one, and vice versa. Consequently, wide ranges of brightness within one's

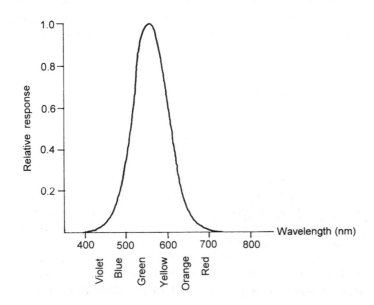

2.6 The spectral response of the average human eye for photopic vision.[17]

field of view are uncomfortable as the eyes do not know which level to accommodate. If they adjust to the higher level, the information from the lower level is lost; and in adjusting to the lower level, the eye gets painfully over-loaded from the higher level.

This is a problem of glare and is illustrated by the classic interrogation technique with the lights facing the victim in an otherwise darkened room. Outside, in daylight, the lights would not be much brighter than their surroundings and the effect would be lost.

Light is directional and travels in straight lines. It is reflected from surfaces: reflection from rough or irregular surfaces is diffuse. Surfaces such as mirrors reflect light directly and this is known as specular reflection.

The colour of a surface determines how much light is reflected and how much is absorbed. White or untinted mirror finishes will reflect almost all the light; black will reflect very little; and shades will vary between these two (Table 8.3). (Reflectance is the ratio of light reflected from a surface to that incident upon it.) The colour of a surface we perceive is the colour of the light hitting the surface less the colours that are absorbed by the surface itself. In white light a green leaf absorbs the blue, red and yellow colours and reflects the green. In the light from a yellow sodium street lamp (which has no green component), a leaf will absorb the yellow and reflect nothing and thus appear black.

Light can reach a surface from a source such as a light bulb or the Sun directly, indirectly by being reflected from surrounding surfaces, or by a combination of both. The colour and reflectance of surfaces do have a dramatic effect on the indirect light component, and therefore the total light level, in a space. The light level in a darkly decorated room will be much less than that in a white-painted room given the same light input. For this reason, to minimize energy consumption it is better to have light or white finishes in spaces.

The ratio of direct to indirect light affects the visual 'feel' and comfort of a space. All direct light with no reflected component (e.g. a matt black room with a point source of light) will show objects in very high contrasts and sharp shadows. At the other extreme a completely diffuse lighting scheme (e.g. an evenly illuminated ceiling with white walls and floor) will be shadowless and without texture.

2.6 Sound

Sound is a waveform like light and is governed by some of the same physical principles. Unlike light, sound is a pressure wave (caused by something vibrating) travelling through a solid, liquid or gaseous medium. It cannot travel through a vacuum. In air it is relatively easy to produce a noise that will travel quite far, and not necessarily in the line of sight. Production of sounds is therefore a useful means of communication that is found throughout the animal world. The management of sounds is an essential part of a building's function to provide privacy, allow intelligible speech, and control of unwanted noise; this is covered by legislation.[18]

The pitch of a sound depends on the frequency of the sound wave and is the equivalent of colour in light – high-pitched sounds are of high frequency. The human voice carries most of its information in the frequency range 125–8000 Hz (cycles per second). The human ear is sensitive to a range of sound frequencies from about 15 Hz, which corresponds to a very low rumble of a distant bus or the lowest organ note, up to 20 000 Hz; door squeaks and the chirp of some insects have a frequency of about 17 000 Hz. However, the ear is less sensitive to high and low frequencies than to those in the middle range.

Sound levels are commonly measured in A-weighted decibels (dBA – see Appendix C for further discussion). This is a scale which takes account of the intensity of all the audible frequencies and weights them in accordance with the ear's sensitivity. It gives a single-valued number that correlates well with the human perception of relative loudness. A sensitive human ear can detect sound down to about 0 dBA. Normal speech is conducted at about 55–70 dBA at the listener's ear. The threshold of pain is approximately 150 dBA.[19] The decibel (dB) is not a linear value but a logarithmic one and a huge range of about 10^{17} is covered by the sounds we hear from the quietest whisper to booster rockets. The ear adapts to different sound intensities and takes time to readjust to changes in levels.

Sound can travel either by a direct path from source to receiver or via an indirect path involving reflections from the surroundings. The perception of sound in an internal environment will result from a combination of these direct and indirect components. The area and absorption of the surfaces in a room will affect the amount of indirect sound and therefore the total sound level within it. It will also affect how reverberant or echoey the space feels. A living room with thickly upholstered furniture, deep carpet, heavy curtains and the odd wall hanging will sound quiet or 'dead', and the volume of the TV will have to be quite high. The same room with no soft furnishings will sound more 'live' and the volume of the TV can be reduced to achieve the same sound level. The overall measure of the absorption within a space is the reverberation time (RT); the time taken for a sound to decay 60 dB.

For music, high reverberation time has the desired effect of blending discrete notes together. The same effect is less appreciated for speech. Discrete words may be blended together, making them unintelligible. There is therefore a conflict between halls used for music, which require high reverberation times, and halls used for speech, which need lower times for intelligibility. Normally, resolution is by designing for a compromise reverberation time between the two extremes. For example, in a room with a volume of around 300 m^3 the optimum reverberation time for speech might be about 0.65 s and, for music, 1.2 s[20] and a value between these two would probably be used.

It has been established that long-term exposure to high noise levels will cause premature hearing degradation, if not loss. EEC guidelines and the Control of Noise at Work Regulations (2005) suggest a daily exposure level limit of 85 dBA.[21] Although, in most situations, noise levels in buildings will be well below this figure, in some cases such as very reverberant canteens and swimming pools, higher levels can be reached and will need to be controlled with absorption.

The ear and brain are very good at filtering sound to extract information such as speech from a background of noise. There are, however, limits and the greater the noise/speech ratio the less information is received. The brain is capable of filling in the gaps in the information to a greater or lesser extent depending on one's prior knowledge of the subject and one's skills of interpolation.

This filtering of speech from noise is particularly noticeable at a party. At the start, when few people have arrived, one can talk normally as the background noise level is low. As more people arrive, the number of people speaking, and therefore the total sound level, goes up. To be heard over the background noise, people must raise their voices and so the noise level rises further. To increase the probability of their speech being understood in spite of the background noise, people get closer and closer together.

It has been found that at a background sound level of 48 dBA the maximum distance for normal speech intelligibility is about 7 m; at 53 dBA the distance falls to about 4 m and so on.[22] Raising the voice increases the distance, and teachers know this well. However prolonged use of the voice at raised levels and the need to shout can lead to problems with voice strain. The DfES make recommendations for the maximum background noise levels (BNL) that are acceptable in a classroom to maintain the teacher's intelligibility to the pupils.[23] For example, the maximum BNL for a large lecture room is given as 30 dBA. Speech intelligibility is also affected by how reverberant the room is and, for this reason, the DfES also set upper limits on room reverberation times in schools.

A room's reverberation time can be tuned to provide the background, diffuse 'burble' from the noise of people and machines in the space. Where there is no noise being produced, such as in a library or waiting room, it needs to be actively added by means of a white noise generator or music in order to provide acoustic privacy by masking.

Sound waves can be reflected or absorbed by surfaces depending on their construction. Generally, hard smooth finishes will reflect sound. Sound can be absorbed in a number of ways, and three common ones in buildings are dissipative (or porous) absorbers, membrane absorbers and cavity absorbers. Here we shall deal only with the first two, and briefly at that. Porous absorbers allow the pressure wave into the surface of the material. Friction between air particles and the material results in dissipation of sound energy as heat. The effectiveness of this absorption depends on the thickness of the absorptive material compared with the wavelength of the incident sound. Thicker materials will absorb longer wavelengths better than thinner ones. Membrane (or panel) absorbers first convert the energy of the pressure wave into vibrational energy in the panel facing, and then further loss occurs in the air space behind the panel. Suspended ceilings and raised floors are two common membrane absorbers. Generally, no single surface provides adequate absorption over a wide frequency range. Membrane absorbers (panels) tend to be better at lower frequencies, and porous absorbers (for example, soft furnishings such as heavy curtains) tend to be better at higher frequencies.

One space can be acoustically separated from another by using solid partitions and by ensuring that no direct air paths connect the two. The heavier the partition, the more difficult it is for the air pressure waves to vibrate it and the greater the separation. This relationship is characterized by the mass law (Figure 2.7). (The sound insulation of an element is basically the difference between the sound level in one room with a noise source and the sound level in an adjacent room that is separated by that element.)

Masonry walls between separate dwellings often have masses of about 380 kg/m^2 and, thus, give sound insulation of about 48 dB. Any cracks forming air gaps through the wall will significantly degrade the sound insulation performance. Un-plastered

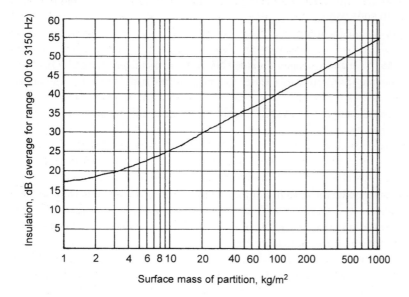

2.7 Sound insulation as a function of surface mass for single-leaf construction.[24]

blockwork is a good example of a construction that is likely to fail due to the presence of air paths.

Double skins of construction separated by an air gap (e.g. stud partitions) can provide greater sound insulation for the same mass of construction. The bigger the air gap, the less the acoustic coupling between the two layers and the better the sound insulation. Absorbent material in the cavity can also improve the performance. It is worth noting though that the sound insulation performance of a lightweight stud partition will be less good at low frequencies than a comparable masonry wall. If not accounted for, this can lead to bass–heavy music passing easily between rooms or dwellings.

Because of the importance of acoustics in buildings we shall return to these issues several times in the following chapters.

2.7 Air quality

Clean, dry air near sea level has the following approximate composition:[25]

Nitrogen (N_2)	78%
Oxygen (O_2)	21%
Argon (Ar)	1%
Carbon dioxide (CO_2)	0.03%

In addition, there are trace amounts of hydrogen, neon, krypton and a number of other gases as well as varying amounts of water vapour and small quantities of solid matter. Because of rising CO_2 production resulting from human activities, the CO_2

level in much of the atmosphere's air is higher, at about 0.037% and increasing by 1.2 ppm per year (0.00012%).[26]

Pollutants in the air include:

- Nitric oxide (NO) and nitrogen dioxide (NO_2), known collectively as NOx
- Sulphur dioxide (SO_2) and, to a lesser extent, sulphur trioxide (SO_3), known as SOx
- Volatile organic compounds (VOC), including benzene and butane
- Carbon monoxide
- Lead
- Ozone.

A number of these pollutants result from the combustion of fossil fuels to meet our energy demands.

In addition, there is a great deal of 'dust' in the air. In fact this is a mixture of dusts, smokes and fumes which are particulate matter and vapours and gases which are non particulate.[27] As an indication of sizes smoke particles are 0.01–0.5 micrometres (μm); mists and fogs are under 100 μm; and pollen is in range of about 10–100 μm. Heavy industrial dust can range from 100–1000 μm and higher.

The amount of solid material in the atmosphere obviously varies enormously, but as a guideline in metropolitan areas it is in the range of 0.1–1.0 mg of material per 1 m^3 of air (mg/m^3) and in rural and suburban areas 0.05–0.5 mg/m^3.[28] Table 2.3 gives a typical analysis for average urban and suburban conditions combined.

There is growing concern about the health effects of most air pollutants and also the combined effect of dust.[30,31] PM10, the name given to particulate matter of diameter 10 μm and less, is now thought to be a cause of cardiovascular and respiratory diseases.

One of the most serious effects of air pollution is 'acid rain', caused chiefly by oxides of sulphur (SOx) and nitrogen (NOx) emitted during fossil fuel combustion and metal smelting.[32] 'Acid rain' is the formation in the atmosphere of acids which are then returned to earth resulting in acidification of streams and lakes, damage to trees and degradation of stonework in historic buildings.

There is an expectation that architects and engineers should be leading the way to design buildings that do not cause pollution, by reducing the energy the building needs and meeting the demands from renewable energy sources.

Table 2.3 Analysis of typical atmospheric dust[29]

Range of particle size diameter (μm)	Percentage of total mass of particles (%)
30–10	28
10–5	52
5–3	11
3–1	6
1–0.5	2
Below 0.5	1

Ventilation is the provision of air to a building. One reason for ventilation is that the occupants need oxygen to oxidize their food (which contains carbon and hydrogen) to produce the energy needed to live. In the process, carbon dioxide and water are formed. The ecological cycle continues with plants taking up carbon dioxide and water in photosynthesis and producing oxygen and food. Our oxygen requirements are met by about 0.03 litre of air per second per person (l/s person).[33]

Removing the carbon dioxide we produce requires higher ventilation rates than those needed to provide oxygen. The occupation exposure limit for CO_2 is 0.5%. Maintaining the room CO_2 level at 0.5% for people involved in light work requires 1.3–2.6 l/s person of fresh air (assuming 0.04% of CO_2 in fresh air); to limit the room CO_2 level at 0.25% requires 2.8–5.6 l/s person.[34]

Ventilation is also required to remove indoor pollutants, to take away moisture to reduce the risk of condensation (Chapter 4) and to take away heat in the summer to maintain comfort (Chapter 9).

Indoor pollutants, some of which originate outside and some inside, include nitrogen dioxide (NO_2), carbon monoxide (CO), carbon dioxide (CO_2), radon, formaldehyde, sulphur dioxide (SO_2), ozone (O_3), mineral fibres, tobacco smoke, body odours and hundreds of other substances.[35,36] Acceptable levels for all of these do not exist and it is not easy to define the quality of air that is suitable. The answer affects health and energy conservation. Ideally, the best approach is to reduce the pollutants at source – a common example at present is to prohibit smoking in many areas. After that there are two approaches: the first is simply to reduce their concentration by diluting them with more 'fresh' air; the second is to filter out contaminants. Both require energy. Museums have very stringent requirements which is something that is reviewed in *Guidelines on Pollution Control in Museum Buildings*.[37]

There is not yet complete agreement on how much fresh air is required in buildings. In the UK, the CIBSE recommends 8 l/s person in offices in the absence of smoking.[38] For schools DfES recommends a minimum of 3 l/s person for naturally ventilated spaces.[39] The figures cited should be 'adequate' to deal with odour control and to provide a reasonable indoor air quality. The pollutants both outside and inside obviously vary, and individual responses are subjective.

The issue of fresh air requirements has been examined carefully in the context of sick building syndrome (SBS). Sick building syndrome was first thought to be primarily a design problem affected by physical features such as the type of ventilation system or the depth of a space; it is now thought that management and maintenance are important.[40] Occupants tend to be more satisfied if the building can respond quickly to requests for change from its users.[41] An aspect of this is a degree of user control as provided by openable windows, adjustable blinds and manually adjustable thermostats.

Occupants make subtle choices between high temperatures, poor air quality and excess noise which are difficult for mechanical systems to match.[42]

Research into health and sick building syndrome has shown that indoor surface pollution – consisting not only of dust but of skin scales, debris from shoes and clothing, the products of smoking, eating and drinking and a number of other sources – seems to be one of the causes of the syndrome.[43]

House mites feed on skin scales and hair. Mites and their faeces are small enough to be suspended in air that is inhaled and can then affect people adversely. Fabrics such as carpets and upholstery are difficult to clean fully and serve as refuges for house mites. The problem of mites and dust is exacerbated by vacuum cleaning. If the cleaner's filter system is not fine enough to remove them, the smallest particles are lifted from the carpet where they are harmless, and are circulated into the room air by the jet of warm air rising from the cleaner's discharge. Solutions offered to this are improving filtration on cleaners, extracting all cleaners' discharges to outside, or, more drastically, removing soft furnishings.

2.8 Moisture

Water and water vapour can cause problems in buildings for two main reasons. Firstly, a number of organisms, whose effects can be detrimental or damaging, can live in buildings and their growth is favoured in humid conditions. Mould, for example, will grow on surfaces that are consistently above 70% r.h.[44]

Secondly, water and high humidity levels can damage the materials used in buildings; for example, exposure of the steel in reinforced concrete to moisture could lead to corrosion and eventual failure.

Sources of water in buildings include rain penetration, rising damp, leaking pipes and the condensation of water vapour in the air. Breathing, washing, cooking and drying all release water vapour.

There are two common mechanisms that remove water vapour from the air in a building:

Dilution with fresh air

If the moisture content of the outside air is less than that inside, ventilation with outside air will dilute and reduce the internal water vapour concentration. Cold air cannot carry as much water vapour as warm air. In the UK and temperate climates the air outside is generally drier than the internal air, even if it is 100% saturated and raining outside. In hot, humid climates high external moisture contents mean dilution with fresh air is not generally possible.

Condensation

If air is cooled, the water-carrying capacity of the air is reduced. At some point the water vapour in the air will reach 100% saturation (the dewpoint) and condense into water (Appendix A). Outside, this commonly occurs at night when the ground loses heat to the night sky and the water vapour in the air condenses as dew. (Air-conditioning systems employ the principle of condensation for dehumidification by passing air over cold metal coils.)

Inside, condensation will form on surfaces that are cooler than the dewpoint of the surrounding air. This can occur in a number of different circumstances. In low-income housing with high occupancies producing a great deal of water vapour and limited money for heating bills, condensation may occur on cold, poorly insulated walls.

In a well-ventilated, well-heated, lightly occupied building such as a stately home, one is more likely to get complaints of the air being too dry and the antiques cracking as a result.

In other buildings the surface of a chilled air duct or a cold water pipe can cause condensation, which can be avoided with a combination of measures:

- reducing the amount of moisture being released within the building;
- removing moisture from the air;
- ensuring that the surfaces are warmer than the dewpoint by a combination of insulating the walls and heating the space.

Preventing condensation at all times can be difficult but may be unnecessary providing that the space has a chance to dry over a daily cycle.[45] For instance, mould will not grow in a bathroom if the condensation due to bathing clears and the moisture content is reduced to below 70% r.h. after use.

Condensation within an element of construction is known as interstitial condensation. It is related to the temperature gradient in the element, say, a wall, which separates a space at one temperature and the outside at another. If water vapour flows through the wall (driven by the difference between the vapour pressures inside and outside) and reaches a part of the construction that is below the dewpoint temperature, condensation will occur. To prevent this, vapour checks are often used in wall constructions to limit the amount of water vapour flowing and to control the vapour pressure distribution in the wall. Repeated and prolonged interstitial condensation can lead to structural problems and must be avoided. Another condensation problem sometimes encountered is due to inadequate air circulation in a room. For example, if a wardrobe is put against a poorly insulated external wall the air between the two will be stagnant and cool. This, in turn, can lead to high relative humidities and condensation, and localized mould growth.

Absorption

Moisture can move through and be stored in materials in an analogous way to heat. An impervious element is equivalent to a thermal insulator and materials that can absorb a lot of water are similar to a thermal mass. This can be used to even out peaks in moisture content in spaces such as bathrooms that suffer very high moisture peaks and in museums where very stable humidity levels are required. In this way the fabric can store moisture when the levels are high and release the water vapour when they are lowered. With this form of construction it is important to check that the moisture can travel through the structure sufficiently quickly to avoid the interstitial condensation discussed above.

Desiccant mechanical systems use absorption to remove moisture from the air in an active manner in air conditioning plants. The advantage of this is that heat from the sun can be used to dry the desiccant as part of the process, rather than using electricity to drive a refrigeration device.

Guidelines

1. Comfort is important for the human body. Designers can increase the likelihood of thermal comfort in their buildings and their acceptability to occupants by providing user control, allowing for ample ventilation and limiting summer-time peak internal temperatures.
2. Both external and internal noise need careful consideration when developing designs.
3. Energy efficient buildings help to improve external air quality and to reduce the effects of global warming.
4. Control of relative humidity and condensation in buildings is essential in building design.

References

1. Anon (2001) *ASHRAE Handbook – Fundamentals*, Chapter 8: Thermal Comfort, ASHRAE, Atlanta.
2. Hughes, J. and Beggs, C. (1986) The dark side of sunlight. *New Scientist*, **111** (1522), 31–5.
3. Pratt, A.W. (1958) Condensation in sheeted roofs. National Buildings Studies Paper No. 23. HMSO, London.
4. Anon. (1988) *CIBSE Guide, Volume A: Thermal Properties of Building Structures*, CIBSE, London.
5. Everett, A. (1975) *Materials*, Batsford, London.
6. Anon. (1993) *ASHRAE Handbook – Fundamentals*, ASHRAE, Atlanta, p. 38.
7. Anon (January 2005) Saint Gobain low-emissivity glass for enhanced thermal insulation. www.saint-gobain.com
8. Anon (January 2005) Pilkington Glass Performance Data Pilkington Insulight™. www.pilkington.com/europe/uk
9. Peuser, F.A., Remmers, K.H., Schnauss, M. (2002) *Solar Thermal Systems: Successful Planning and Construction*, Solarpraxis AG, Germany.
10. Ibid.
11. Anon. (2003) Buildings Bulletin 87. 2nd Edition Version 1, Guidelines for Environmental Design in Schools. Department for Education and Skills, London.
12. Taylor, A. (1987) Curing windows pains. *Energy in Buildings*, **6**(6), 21–4.
13. Moon, P. (1940) Proposed standard solar radiation curve for engineering use. *Journal of Franklin Institute*, November, p. 604.
14. Anon. (1987) *ASHRAE Guide: HVAC Systems and Applications*, ASHRAE, Atlanta.
15. Anon. (2004) Climate change: Action to tackle global warming. DEFRA. www.defra.gov.uk
16. Anon. (January 2005) Emissions inventories based on the methodology developed by the Intergovernmental Panel of Climate Change (IPCC) used to report UK emissions to the Climate Change Convention. DEFRA, London.
17. Collingbourne, R.H. (1966) General principles of radiation meteorology, in *Light as an Ecological Factor* (eds R. Bainbridge, G.C. Evans and O. Rackham), Blackwell, Oxford.
18. Anon. (1981) *Guidelines for Environmental Design and Fuel Conservation in Educational Buildings*, Department of Education and Science, London.
19. See reference 11, p. 5.
20. Sharland, I. (2001) *Woods Practical Guide to Noise Control*, Woods Air Movement Limited, UK.
21. Anon. (1986) EEC Council directive 86/188 on the protection of workers from the risks related to exposure to noise at work. *Official Journal of the European Communities*, 12 May 1986. HMSO, UK.

22. Anon. (1988) *CIBSE Guide A1: Environmental Criteria for Design*, CIBSE, London.
23. Anon. (1975) *Acoustics in Educational Buildings*, Department of Education and Science, Bulletin 51. HMSO, London.
24. Anon (1988) Insulation against external noise. BRE Digest 338. BRE, Garston.
25. Anon. (2001) *ASHRAE Handbook – Fundamentals*, Chapter 12: Air Contaminants, ASHRAE, Atlanta.
26. Anon. (2001) Contribution of Working Group 1 to the Third Assessment Report of the Intergovernmental Panel on Climate Change (IPCC), Cambridge University Press, Cambridge.
27. Anon. (2001) *CIBSE Guide B2: Ventilation and Air Conditioning*, CIBSE, London.
28. Ibid.
29. Ibid., chapter 5–14.
30. Read, R. and Read, C. (1991) Breathing can be hazardous to your health. *New Scientist*, **129**(1757), 34–7.
31. Bown, W. (1994) Dying from too much dust. *New Scientist*, **141**(1916), 12–13.
32. Gorham, E. (1994) Neutralizing acid rain. *Nature*, **367**(6461), 321.
33. Mayo, A.M. and Nolan, J.P. (1964) Bioengineering and Bioinstrumentation, in *Bioastronautics* (ed. K.E. Schaeffer), Macmillan, New York.
34. Anon. (1991) Code of practice for ventilation principles and designing for natural ventilation. BS 5925 : 1991. British Standards Institution, London.
35. See reference 25.
36. Anon. (2001) *ASHRAE Handbook – Fundamentals*, Chapter 9: Indoor Environmental Health, ASHRAE, Atlanta.
37. Blades, N., Oreszczyn, T., Bordass, B., Cassar, M., (2002) *Guidelines on Pollution Control in Museum Buildings*, Museum Practice, London.
38. See reference 22, pp. A1–9.
39. See reference 11.
40. Leaman, A. (1994) Complexity and manageability: pointers from a decade of research on building occupants. *Proceedings of the National Conference of the Facility Management Association of Australia, Sydney*.
41. Ibid.
42. Anon. (1993) What causes discomfort? *Buildings Services*, **15**(6), 47–8.
43. Raw, G.J. (1994) The importance of indoor surface pollution in sick building syndrome. BRE Information Paper 3/94. BRE, Garston.
44. Anon. (1985) Surface condensation and mould growth in traditionally built buildings. BRE Digest 297. BRE, Garston.
45. Ibid.

Further reading

Addleston, L. and Rice, C. (1991) *Performance of Materials in Buildings*, Butterworth-Heinemann, Oxford.

Anon. (1988) Sound Insulation: basic principles. BRE Digest 337. BRE, Garston.

Anon. (2003) *ASHRAE Handbook – Fundamentals*, Chapter 8: Physiological principles and thermal comfort, ASHRAE, Atlanta.

Anon. (2003) *ASHRAE Handbook – Fundamentals*, Chapter 9: Indoor environmental health, ASHRAE, Atlanta.

Anon. (2003) *ASHRAE Handbook – Fundamentals*, Chapter 12: Air contaminants, ASHRAE, Atlanta.

Bartlett, P.B. and Prior, J.J. (1971) The environmental impact of buildings. BRE Information Paper 19/91. BRE, Garston.

Evans, B. (1994) Making buildings healthier. *Architects's Journal*, **202**(23), 57–9.

Fry, A. (ed.) (1988) *Noise Control in Building Services*, Pergamon, Oxford.

Getz, P. (ed.) (1986) *The New Encyclopedia Britannica*, Encyclopedia Britannica, Chicago. Selected articles on convection, heat transfer, etc.

Gribbin, J. (1988) The ozone layer. *New Scientist*, **118**(1611), 1–4.

Gribbin, J. (1988) The greenhouse effect. *New Scientist*, **120**(1635), 1–4.

Gribbin, J. (1988) Quantum Rules, OK!. *New Scientist*, **123**(1682), 1–4.

Rennie, D. and Parand, F. (1996) Environmental design guide for naturally ventilated and daylit offices, BRE paper BR345, BRE, Garston.

Buildings and energy balances CHAPTER 3

3.1 Introduction

In this chapter the importance of buildings to energy use and carbon dioxide production in the UK is indicated. Energy use in different types of buildings is also examined in preparation for reducing it through design in subsequent chapters.

3.2 Buildings in the broad context

Buildings use energy and, as most of the UK's energy comes from the combustion of fossil fuels, produce CO_2 in the process. Delivered energy is the energy in fuels at their point of use (Chapter 7). Figure 3.1 shows fuel consumption and carbon dioxide emissions in the UK. Buildings account for about 45–50% of delivered energy use and just under 50% of all CO_2 emissions. The UK itself is responsible for about 3% of global CO_2 emissions.

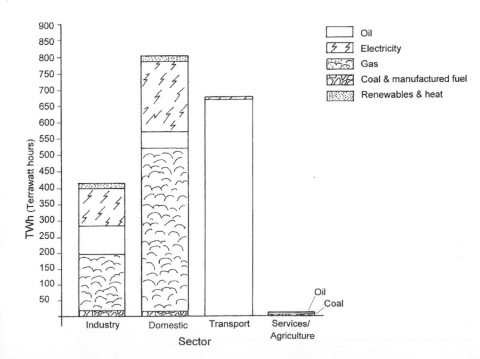

3.1(a) UK fuel consumption by sector and by fuel type (2003).[2]

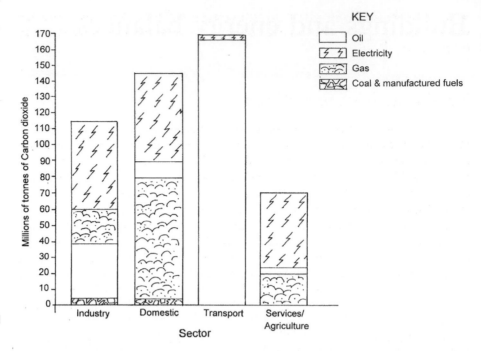

3.1(b) UK carbon dioxide emissions by fuel and by sector (2004).[3]

KEY

- Oil
- Electricity
- Gas
- Coal & manufactured fuels

Approximately 60% of building-related CO_2 emissions is due to the domestic sector and about 30% is attributable to the service sector, i.e. UK public and commercial buildings.[1] In the service sector the total CO_2 emission is about 89 million tonnes and approximately 44% of this is due to space heating (Table 3.1). Architects can have a major input through design on reducing these figures.

While architects can influence new building design fairly easily, it must be recognized that the vast majority of the UK's building stock is already in existence. To reduce energy consumption the insulation of these buildings needs to be upgraded, but in practice this can be difficult. Increasing the loft insulation of housing has proved to be one of the easier approaches so far. Additionally, replacing windows

Table 3.1 UK carbon dioxide emissions by end use for the UK service sector[4]

Use of fuel	%
Space heating	44
Water heating	7
Lighting	17
Cooking	6
Air conditioning	6
Refrigeration	7
Power	13

with better sealed double glazed units, insulating hot water cylinders, installing condensing boilers and better controls contribute to a reduction in domestic energy consumption.

3.3 Energy flows in buildings

In Chapter 2 we saw that energy exchanges affect people and buildings. Externally, the building envelope is subject to solar gain, radiation exchange with its surroundings and convective heat loss (or gain) owing to the winds that almost continuously flow past it. Moisture, too, can be lost to the surroundings (and, somewhat differently from the human body, gained as when driving rain penetrates into the external wall). Internally, the building is the site of the activities of the occupants and processes which include lighting, running of equipment from ventilation fans to photocopiers, cooking and heating. Many of the internal loads are appliances that are outside the control of the designer. All of the energy supplied finishes as heat, even when there is a useful intermediate state such as light or a moving fan. There is now pressure being brought to bear on producers to label the energy consumption of appliances to enable the consumer to make educated choices about lower energy appliances (see Chapter 10).

Table 3.2 shows some typical energy consumption figures for a variety of processes and the associated CO_2 production.

Figures 3.2 and 3.3 show typical energy flows in a house and an air-conditioned office.

If we first consider the domestic situation, we see in Figure 3.2 that there is an input from the Sun which contributes to heating and lighting and which, although free, is highly valuable.

The advantages of daylight and solar gains need to be balanced against potential overheating and heat loss through glass. This issue also obviously applies to offices and other building types. It illustrates the need to view the building as a system and at times to balance contradictory factors. If one element is given too much emphasis – for example, daylighting in system-built schools in the UK in the 1960s – problems can result. In the summer such schools suffered from overheating and glare, and in the winter they had high heat losses. If, on the other hand, form, glazing, fabric, ventilation and services work together, comfort and energy efficiency can be achieved.

Many loads in buildings are highly intermittent, such as hot water use for showers or washing hands, or energy use for cooking. Others, such as heating and lighting, can be more constant.

Loads need to be examined and quantified so as to be able to identify how best to provide the energy required and how to reduce the demand as much as possible. Lighting levels in a domestic living room are likely to be much lower than in offices, but because the office lights tend to be much more efficient the actual electricity consumption may be similar.

In a domestic situation about 23% of the total energy requirement is for heating water (for handwashing, bathing, cooking and so forth).[7] Much of the heat is lost down the drains or goes into evaporating water which is added to the internal air.

Table 3.2 Approximate energy consumption and carbon dioxide production for selected activities, equipment and buildings[a]

Item	Energy consumption (kWh)	CO_2 production (kg)	Period	Notes
Man at rest	2.8		Day	
Shower	1.8	0.3	5 minutes	Water heated with gas
Bath	3.3	0.6		80 litre bath heated with gas
Dishwasher	2	1.0	1 cycle	Including heating the water with electricity
Fridge/freezer	2.2	1.1	24 hours	
100-watt filament light bulb	2.4	1.3	24 hours	
Equivalent miniature fluorescent	0.5	0.3	24 hours	
For an average 3-bedroom dwelling				
Electricity consumption	2 000	1 040	1 year	Excluding water or space heating
Water heating	2 000	380	1 year	
Space heating – Old housing	22 500	4 280	1 year	Using gas
Space heating – new house target	3 500	670	7 year	Using gas; 100 m^3
Domestic heat recovery ventilation system	2 625	560	1 year	Electricity in running the fans; 100 m^3
For a 1500 m^2 primary school				
Electricity use	27 000	14 040	1 year	
Gas use	75 000	14 250	1 year	Top 10%[5]
For a 15,000 m^2 air-conditioned office				
Electricity use	351 000	182 500	1 year	Good practice, see
Gas use	171 000	32 500	1 year	Figure 3.4
100 km car journey	100	22	2 hours	
1 acre of corn		3 000	1 year	CO_2 fixed

[a] Figures are indicative and can often be improved with design.

Ventilation provides oxygen, fresh air, removes CO_2 and odours and helps prevent condensation problems by taking away water vapour. Heating systems replace heat loss through the fabric (which is falling as insulation standards increase) and heat the ventilation air in the winter.

The heat requirement for the ventilation air supply becomes more and more important as fabric heat losses fall.

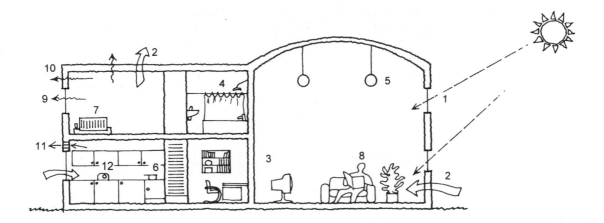

Key
1. Solar gain 200–600 W/m^2 of windows
2. Fresh air 10 l/s 0 °C to 21 °C = 250 W
3. TV 100–400 W
4. Shower 1.5 l/s, ΔT = 30 °C = 18 900 W
5. Lighting: domestic lamps are 60–100 W, small fluorescents are 11–17 W
6. Cooker rings 500–1500 W. Most of the heat goes into turning water into steam.
   Oven 1000–3000 W
7. Radiator 1000 mm long x 490 mm high (single panel): 670 W
8. People 80–300 W (higher value when exercising)
9. 1m^2 window heat loss
   110 W/m^2 single glazing
   50 W/m^2 double glazing
   40 W/m^2 double glazing low-emissivity coating
   Assuming a 20 K temperature difference between inside and out
10. Building fabric
   40 W/m^2 if solid brick
   10 W/m^2 wall with 100 mm of insulation
   5 W/m^2 wall with 150 mm of insulation
   Assuming a 20 K temperature difference between inside and out
11. Extract fan 60–120 W
12. Radio 4 W

3.2 Approximate energy flows in a house (Sunday) at home.

As the heat retention and recovery systems improve, the internal heat gains begin to become sufficient to heat the buildings for some or all of the winter.

These same gains can also create more of a problem in summer and potentially cause overheating. The only passive way of getting rid of the heat is by ventilation. The fabric can store the excess heat during the day which can then be purged with cooler night air. (Chapter 18 describes the natural ventilation and heat extraction systems at Heelis, the National Trust Offices.)

Most air-conditioned buildings are completely sealed off from the outside and air is supplied from a central air-handling plant (shown schematically in Figure 3.3). Heat

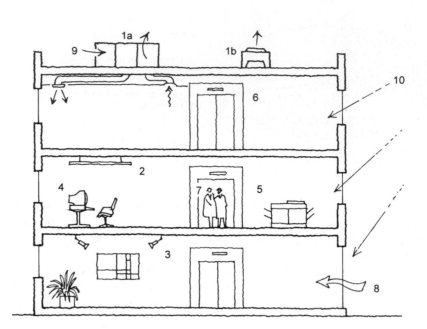

3.3 Approximate energy flows in an air-conditioned office with summer cooling loads.

Key
1. Air conditioning
   (a) air-handling unit
   (b) heat rejection equipment
2. Fluorescent lighting
3. Tungsten lighting
4. Computer
5. Photocopier
6. Lift
7. Person
8. Warm air entering in summer through open door
9. Fresh air
10. Solar gain

10–30 W/m² of office for total plant electrical load including chiller (not shown)

10–20 W/m² of office to give 300–500 lux
20–100 W/m² of office to give 100–300 lux
50–250 W
1000–5000 W
10 000 – 30 000 W
100–140 W
2000 W (1 m³/s)

150 W per person at 10 l/s person
200–600 W/m² of window

from the Sun, equipment and people must be removed by air from the central plant and this necessitates a cool air supply and so requires energy. Figure 3.4 gives typical data.

Energy for cooling (and associated equipment) and lighting are two areas where significant reductions can be made, and we shall examine some approaches in later chapters.

Construction in theory and practice

There is an expectation that buildings will be better insulated and better sealed than they have in the past. This increasing standard will be challenging for all professionals and constructors in the industry who will have to learn new ways of working.

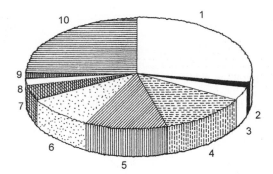

KEY

1. Gas/oil heating and hot water (30%)
2. Catering gas (2%)
3. Cooling (6%)
4. Fans, pumps and controls (10%)
5. Humidification (6%)
6. Lighting (8%)
7. Office equipment (6%)
8. Catering equipment (4%)
9. Other electricity (4%)
10. Computer room (24%)

3.4 Delivered energy use in a typical air-conditioned office building.[6]

Designers will have to understand how to seal and insulate a building and understand the limitations of the constructor's capabilities. Poor design will simply not work, whilst badly communicated design runs the risk of being installed incorrectly without adequate supervision.

Energy savings can also be made with better control of the heating, cooling, ventilation, lighting etc. Again the industry is used to making the systems work to prevent complaints for the occupants, not to save energy. This will be another cultural shift.

The new building regulations in the UK require very high standards of construction and testing of buildings for air tightness. These standards are well ahead of the construction industry's capabilities currently.

Well-sealed buildings will bring other problems of air quality, moisture and mould growth which must be attended to with careful control of the ventilation.

Buildings operation and energy monitoring

Building services systems should be well run and controlled to minimize the energy consumption. However the usage pattern or control systems may subtly alter or fall out of calibration over time and result in high-energy consumption. To ensure that this is not missed, each system should have sufficient sub meters installed to pin point changes in consumption. This allows the management and the maintenance regimes to check why a system might be using more energy.

Energy labelling

One method of using consumer power to reduce energy usage is the government requirement to label products with their energy usage. This has been done for household appliances and is being discussed for buildings.

Guidelines

1. Buildings are major consumers of energy and producers of CO_2. Architects and engineers can help change this.
2. Analysis of energy use and flows helps identify where savings can be made.

References

1. Shorrock, L. (1994) Private communication.
2. Anon. (2004) UK Energy in Brief July 2004. DTI, London.
3. Ibid.
4. Moss, S. A. (1994) Energy consumption in public and commercial buildings. BRE Information Paper 16/94. BRE, Garston.
5. DFES National Stats Energy and Water benchmarks for maintained schools in England.
6. Anon. (2004) *Energy Efficiency in Buildings, CIBSE Guide F*, CIBSE, London.
7. Taylor, L. (1993) *Energy Efficient Homes: A Guide for Housing Professionals*, Association for the Conservation of Energy and the Institute of Housing, London.

Building planning and design CHAPTER 4

4.1 Introduction

This chapter covers shape and size, the 'body' and the 'skin' of the building and issues of internal organization. It provides a basis for articulating the building on the site in order to provide an energy efficient and comfortable internal environment.

4.2 Form

The orientation of a building may be fixed but if choice is possible it should face south to take advantage of the Sun's energy (Chapter 5). Total volume, too, is likely to be prescribed and so, often, the first major design decisions are allocating volumes to various activities and developing the form of a building.

Form is governed by a number of functional considerations that are discussed below, and in more detail in the following chapters, and include:

- the use of the Sun's energy and daylight (Chapters 5 and 8)
- provision of views for occupants
- heat loss through the building envelope
- the need for ventilation (Chapter 9)
- acoustic attenuation if required.

In the recent past, the glass blocks of Mies Van der Rohe epitomized an architecture that shut out the natural environment and provided an acceptable internal environment through the use of considerable energy and sophisticated services.

The Queens Building at De Montfort (Chapter 12) is the antithesis of this and articulates the building both on plan and in section to respond to the environment and make the best use of natural energy sources. The likelihood is that environmental considerations will allow for freer forms and, thus, a welcome architectural diversity; but before we can draw any conclusions about form we need to know more about how buildings work.

4.3 The building 'body'

An important consideration is how quickly a building responds to heat inputs (internal and external), and this is related to the thermal conductivity of its materials, the thermal mass (or heat capacity – both discussed above in Chapter 2), and, related to these, the admittances of the elements of the construction.

The admittance, Y, of a constructional element, put simply, is the amount of energy entering the surface of the element for each degree of temperature change just outside the surface and, as such, has the same units as the U-value (W/m^2 K) (Appendix B). The admittance of a material depends on its thickness, conductivity, density, specific heat and the frequency at which heat is put into it. (In addition to the admittance, the response of building elements to energy cycles depends on the decrement factor and the surface factor;[1] put simply, once again these factors are associated with time lags in energy flows, with the decrement factor representing the 'damping' effect of an element's response to an energy gain.) Considerably more technical explanations of these concepts are to be found in references 1, 2 and 3. Table 4.1 gives properties of some constructions.

As can be seen from the table, dense constructions have higher admittances, which is to say they absorb more energy for a given change in temperature. (One must be careful, however, because for multilayer slabs, the admittance is determined primarily by the surface layer; thus, a 300 mm slab with 25 mm of surface insulation will respond more as a lightweight than as a heavyweight material).[4]

If a building absorbs a great deal of heat and only experiences a small temperature rise it is said, in no very precise manner, to be thermally heavyweight. Such buildings tend to have high admittances and a great deal of thermal mass, usually in the form of exposed masonry. Lightweight buildings, on the other hand, may have thin-skinned walls, false ceilings with lightweight panels, metal partitions and so forth. (A more precise definition of 'lightweight' and 'heavyweight' structures in terms of admittance values, U-values and the ventilation conductance can be found in reference 5.)

The particular importance of these issues is in providing comfortable conditions in the summer without the use of air conditioning. This is not simply a problem for office buildings – countless schoolchildren in the UK were educated in the 1960s, 1970s and 1980s in lightweight, underinsulated, overglazed buildings that overheated in the summer, particularly on the top floor in westerly-facing classrooms in the late afternoon.

Table 4.1 Admittance and density of selected construction elements[6]

Item	Admittance (W/m^2 K)	Density (kg/m^3)
1. 220 mm solid brickwork, unplastered	4.6	1700
2. 335 mm solid brickwork, unplastered	4.7	1700
3. 220 mm solid brickwork with 16 mm lightweight plaster	3.4	1700 for brickwork 600 for plaster
4. 200 mm solid cast concrete	5.4	2100
5. 75 mm lightweight concrete block with 15 mm dense plaster on both sides	1.2	600 for concrete 600 for plaster

Normally, the heat flow into a building from the outside is approximately cyclical. On a daily basis, the Sun rises, the air temperature increases and heat is transferred directly via windows and indirectly via the building structure. As the Sun sets the building starts to cool, and the following day the cycle continues. In the winter, the external gains are insufficient and so the heating system supplies heat each day during the period of occupancy. At night, the temperature is allowed to drop to conserve energy. Again, the following day the cycle continues.

The thermal mass of the building evens out the variations. In the summer, by delaying the transfer of heat into a building, the time the peak temperature is reached can be altered. By using high-admittance elements the building fabric can store more of the heat that reaches the internal and external surfaces, thus reducing the peak temperatures. This 'balancing' effect can apply both during the day and at night, because if cool night air is brought into contact with high-admittance surfaces their temperatures will drop, i.e. there will be cool thermal storage. The next day, when warmer day-time air flows over the same elements, they will be cooled thus improving comfort conditions for the occupants. This technique is used both at RMC (Chapter 11) and De Montfort (Chapter 12).

Architecturally, the key requirement is to incorporate high-admittance materials in the building and expose them in an appropriate manner. This means that false ceilings, raised floors and plastered walls will need to be kept to a minimum. Appearance will obviously be an important consideration as walls and soffits (and services) are bared. However, there are a number of solutions – from brickwork walls with coloured bands to make them more attractive, to high-admittance ceiling linings such as cement-bonded chipboard. It may also be possible to exploit more complex approaches such as taking the incoming air supply over a concrete floor slab. Greater floor-to-ceiling heights will also, of course, provide more thermal mass for a given floor area. In some cases one element may be made to perform several functions. At De Montfort the heavy masonry stacks ventilate, provide thermal mass and help support the roof.

Heavyweight buildings have an important role to play where air-conditioning might otherwise be needed. However, study of a number of buildings has shown that:

–   if loads are low, there is a limit to the need for thermal mass, and
–   there can be a limit to its usefulness.[7,8]

To make efficient use of mass, one must be able to ventilate at night to lower the temperature, otherwise the heat absorbed tends to accumulate and discomfort results. This has practical implications: if night-time ventilation is under automatic control, the system should not be too complex; if under manual control, it needs to be foolproof, both in maintaining security and in preventing the entry of rain.

If loadings are low and air movement is good, comfort can be achieved with light-weight buildings. If a building is always in use – for example, sheltered housing schemes – heavyweight buildings are often appropriate, but if occupancy is intermittent a lightweight building can have a positive advantage. For example, in winter, the heat stored in a heavyweight building during the day may be released at night when there is no need for it. The process is somewhat similar to an electric storage heater that supplies heat during the day when needed, but cannot stop releasing heat after people have left. The significance of this, however, depends on the building; and as insulation standards increase and buildings become better sealed, there is a decrease in the amount of heat wasted by a building when all the occupants have left.

Unfortunately, there are no definite rules; each building needs to be examined on its own merits, and we shall return briefly to these considerations later in this chapter.

4.4 The building 'skin'

Development of the building envelope, or 'skin', is likely to be rapid in the next decade or so. Technological innovation in glass will allow window systems to respond to environmental conditions in ways not previously commercially viable for buildings. Sun-glasses which react to different light conditions are but a hint of the potential of glass.

Building envelopes obviously need to be durable, economical, aesthetically pleasing, weathertight, structurally sound and secure. Psychologically, views out are very important. Environmentally, the questions that need to be addressed are: how they respond to solar radiation (both for the Sun's heat and light), how ventilation is made possible, how heat loss is minimized and how noise is controlled. The envelope will, to a large extent, determine how the internal environment is affected by the external one.

Solar radiation

Figure 2.4 shows the spectral distribution of solar radiation, to which the components of the envelope react in different ways. If we first consider the opaque elements, the amount of radiation absorbed at the surface depends in part on the colour of the surface. Lighter colours, of course, absorb less and reflect more of the incident radiation (Table 2.2).

Turning to translucent materials each one has a different characteristic. Figure 4.1 shows the energy exchange for plain 3 mm float glass. The percentage of solar radiation transmitted by a window varies with wavelength, as shown in Figure 4.2.

Figure 4.2 shows that glass filters the Sun's radiation much as the atmosphere does, absorbing some of the UV and infrared and letting through much of the visible light. A glasshouse will let in a great deal of solar radiation but will not transmit much of the far infrared produced by the room, much as clouds block the Earth's outgoing radiation. (See Figure 2.3 for an approximate spectrum of room radiation. Figure 4.2 does not continue far enough to the right to show the reduced transmission of clear float glass at longer wavelengths.)

The amount of radiation that enters and exits a room can be controlled to a certain extent by altering the components of the glass, by using several layers of glazing, by applying special coatings and filling the spaces between the panes with various gases, or by evacuating them; an example of the altered transmission characteristics is seen in the graph in Figure 4.2.

The heat loss from any building element is related to its U-value (Appendix B). U-values for different glazing types along with transmission and acoustic characteristics

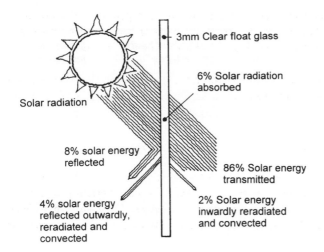

3mm Clear float glass

6% Solar radiation absorbed

Solar radiation

8% solar energy reflected

86% Solar energy transmitted

4% solar energy reflected outwardly, reradiated and convected

2% Solar energy inwardly reradiated and convected

4.1 Energy exchange at a window of 3 mm float glass.[9]

are shown in Table 4.2. (Note that this is for glazing alone; a more precise analysis would be needed to take the frame into account.)

The table shows that there is some loss of light and solar radiant heat as the U-value improves. However, in most applications this is not a significant disadvantage compared with the benefits obtained. It also shows that direct solar transmittance is not the same as direct light transmittance, and this suggests possibilities for glass development. In the summer, for example, an ideal glass would transmit light (to reduce the need for artificial lighting) but no other part of the solar spectrum (to keep the space cooler). In the winter both light and heat are likely to be advantageous. Similarly, in the winter a very low U-value saves energy. If, in the summer, the internal temperature is above the external – as often occurs in lightweight, non-air-conditioned buildings – a high U-value would help get rid of the heat. Glasses whose characteristics can be altered have enormous potential.

Energy loss through a window depends particularly on internal and external temperatures and is independent of orientation. Energy gain, on the other hand, obviously depends on direction because of the Sun. Appendix A gives a selection of

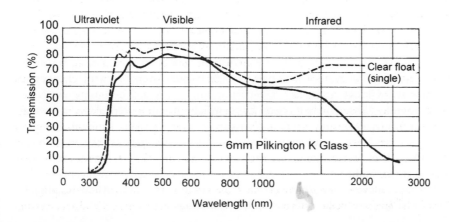

4.2 Spectral transmission curves for glass.[10]

Table 4.2 Characteristics of glazing systems [11]

Type	U-value (W/m² K)	Light transmittance	Solar radiant heat transmittance [a]		Mean sound insulation [b]
			Direct	Total	
Single (4 mm Pilkington Optiwhite clear float glass)	5.8	0.92	0.91	0.91	27
Double glazing (6 mm Optifloat clear float inner, 16 mm airspace,[c] 6 mm clear float outer)	2.7	0.80	0.64	0.72	30
Double with low emissivity coating and cavity (6 mm Pilkington K inner, 16 mm airspace with argon, 6 mm Optiwhite clear float outer)	1.5	0.76	0.61	0.77	30
Triple glazing (4 mm/12 mm/4 mm/12 mm/4 mm) (low emissivity coating and argon filled)	1.0–1.5[d]				

[a] Direct solar radiant heat transmittance covers the entire solar spectrum of approximately 300–2500 nm. Total solar radiant heat transmittance is the sum of the direct transmittance plus the proportion of absorbed radiation re-radiated inwards (reference 12).
[b] Mean sound insulation is for the frequency range of 100–4000 Hz.
[c] At airspaces above 12 mm the U-value is about the same. Below 12 mm it gets worse and typical U-values for 6 mm and 3 mm gaps are 3.2 and 3.6 W/m² K, respectively.[13]
[d] Variable. Consult manufacturers' literature, e.g. see reference 14.

solar data. When solar radiation data is used with internal and external temperatures it is possible to determine the daily solar heat gain and average conduction heat loss – the difference between the two is the daily energy balance. (Figure 4.3 shows energy balance data.)

Newer glasses are likely to have even more favourable energy balances but Figure 4.3 nonetheless presents a broad picture that energy is available and that we should be using it. In doing so we shall, of course, need to guard against overheating and glare. We shall return to this when we discuss shading devices.

First, however, it is worth while just mentioning some areas of glazing research. Photochronic (light-activated) materials change colour reversibly with changes in light intensity. Glasses and plastics incorporating them are becoming available. Thermochromic glasses include clear films which, when heated above a certain temperature, turn an opaque white and so can be used to reflect sunlight. Electrochromic glass has layers whose properties change when a voltage is applied to them. In this way parts of the solar spectrum can be reflected from the glazing thus reducing the heat that enters the building. Transparent (or more precisely, translucent) insulation materials include glass/aerogel/glass assemblies and polycarbonate plastics which can be applied to the external walls of buildings.

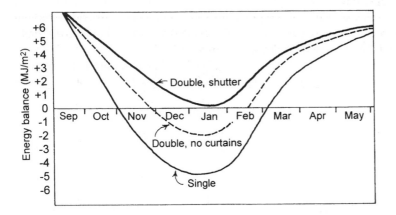

4.3 Daily energy balance for south-facing glazing at Bracknell, Berkshire; assumed internal temperature 18 °C.[15]

Shading is normally needed to control overheating in the summer and in some designs in the spring and autumn. Shading devices need to be considered as systems rather than isolated elements and shading control at the building envelope must be related to the activities in the building, its mass and its ventilation system. A historical example, illustrating the need for a systems view, is Le Corbusier's Salvation Army Hotel in Paris. The original design included a way of removing heat from in front of an inner skin of unopenable south-facing glazing. However, for cost reasons the design was altered leaving only the fixed glazing which almost roasted the occupants.[16] Later, a brise-soleil, or sun screen, was added to reduce overheating.

A disadvantage of external, fixed shading is that it results in some permanent loss of passive solar gain when needed. Note that this is true even if one attempts to design an arrangement based on direct solar gain which blocks out the Sun's rays in June but allows them entry in December. There is also a permanent loss of daylight with fixed external shading. Nonetheless, architects are often drawn to it because it can enliven an otherwise banal façade.

It is, nevertheless, too easy to say 'avoid external shading' and it is much better to examine the functional requirements. Structural overhangs are one form of shading that also offers the possibility of rain-shielding, as shown in Figure 4.4, which is based on the Regional Museum of Prehistory at Orgnac l'Aven in France.

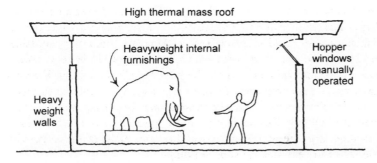

4.4 External shading and night-time ventilation.

In this situation the required light levels are low. A combination of overhang and hopper window means that very simple night ventilation can be provided without major risk from rain entering if someone forgets to close the windows; the night ventilation works well with the high thermal mass.

Movable external devices tend to be costly and because of exposure to the weather require significant maintenance. They should be used only after careful consideration.

Movable shading devices such as blinds placed between glazing layers let more heat into a space than external shades but are more reliable. They are also more effective than internal blinds (but more costly). Very approximately, if single glazing allows in 1.0 unit of solar energy, single glazing with internal blinds will allow 0.67 unit. Double glazing will allow 0.88 unit and double glazing with internal blinds, 0.33 unit.

Figure 4.5 shows 25 mm blinds between two panels of sliding glass at St John's College, Cambridge.

Internal shades include curtains, blinds (with wooden, metal or plastic slats) and shutters. Shutters may, of course, include slats. Curtains, blinds and shutters may incorporate thermal insulation with options varying from simple slabs of, say, mineral fibre built into shutters, to sophisticated aluminium and polyester layers in blinds.

Increasingly, internal shading devices also need to be considered in connection with ventilation (Chapter 9) and in some cases, such as lecture theatres and classrooms, the need to black out (or more precisely, grey out) a space.

4.5 Venetian blinds in the cavity of a double window.

Many of the considerations above apply equally well to daylight (Chapter 8). A résumé of shading devices is given in reference 17.

The world is littered with inappropriate, dysfunctional, flimsy, unreachable, unmaintainable, unaesthetic shading devices. Beware!

Ventilation

Ventilation of buildings has varied from uncontrolled infiltration at cracks around windows, doors, junctions, floorboards and so forth for, say, homes, to purpose-made openings to provide air to air handling units in sealed air-conditioned offices. In most domestic situations it was assumed (usually correctly) that enough air would enter the rooms to meet the needs of oxygen, odour and pollutant removal, condensation control and, in the summer, possible removal of heat. Indeed, generally too much air gained entry during the winter and this led to excessive heat loss. Other problems also occurred but they tended to be localized, and were frequently the result of a combination of high moisture production, low temperatures and inadequate ventilation. The resultant condensation and attendant problems have been mentioned in Chapter 2.

For several years, partly as a result of increased interest in energy conservation, there has been a growing interest in ventilation, summarized by the slogan 'Build tight, ventilate right'. If the right amount of air is provided, if heating systems distribute heat as needed throughout the building, and if moisture is dealt with at the source, for example through extract fans in kitchens and bathrooms, there is a high probability that condensation will be controlled and energy consumption kept reasonably low.

But how to provide the 'right' amount of air? As a starting point, the building needs to be tightly sealed (the adage is 'build tight, ventilate right') so that entry and exit points for air are controllable, or at least well defined. A tightly sealed construction requires careful design and good workmanship. Flexible sealants are required at junctions, say, between window frame and walls and at interfaces of steel frames and masonry; and when detailing external joints, allowance must be made for thermal expansion, deterioration, distortion and weathering. It is not uncommon now to pressure test buildings to ensure that they meet standards of air tightness; Scandinavian buildings often have leakage rates below 5 m^3/h per m^2 of building envelope area. Openings for ventilation will vary according to the application, and windows have normally been the main means of providing natural ventilation in both winter and summer. (Different types are discussed in Chapter 9.)

For small amounts of (permanent, not fully controllable) winter ventilation trickle ventilators incorporated into window frames have become popular, especially in domestic situations; the example shown in Figure 4.6 incorporates acoustic attenuation.

Ventilators are also available which incorporate temperature and humidity sensors and open at the set points, thus providing direct control.

Heat loss

Heat loss at the building envelope is principally a matter of the U-values of the glazed areas (Table 4.2) and the insulation used in the construction of opaque wall elements, roofs and (to a lesser extent) floor slabs. Insulants will be discussed in Chapter 6.

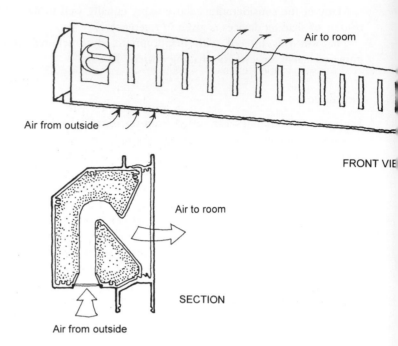

4.6 Trickle ventilators.

Noise

Control of noise is a key issue for many sites, and in many ways noise and acceptable air quality create more difficult problems than overheating due to solar gains. Progress in developing successful designs will have major implications for eliminating the need for air-conditioning and elaborate mechanical ventilation systems. Appendix C gives some of the basic terms and data needed to discuss the issues.

To give an idea of the scope of the problem in urban situations, it has been suggested that, for design purposes, an L_{10} value of 77 dBA be used for noise outside office and commercial buildings.[18] If we try to design for, say, an L_{10} value of 45 dBA inside we would need 32 dB of attenuation. Now let us consider the building envelope as being made up of the opaque solid elements, the glazing and the ventilation openings. Typical masonry walls (and many other kinds) have no difficulty (Figure 2.7) meeting this attenuation figure as long as they are well constructed, but double-glazing systems do not quite meet the requirement (Table 4.2). (To achieve very high values of attenuation it is necessary to increase the space between the panes significantly. A typical acoustic glazing unit might consist of 6 mm glass, a gap of 200 mm or more and a sheet of 10 mm glass with the panes of glass not parallel to each other and absorbent reveals; this could provide up to 45 dB attenuation.)

The weakest link, however, lies with openings for ventilation because, generally, where air can enter, noise can also enter. An open window has an average attenuation of about 10 dB.[19] An opening in a wall can incorporate acoustic attenuation in a variety of ways and the performance will vary accordingly. Figure 4.7 shows a typical

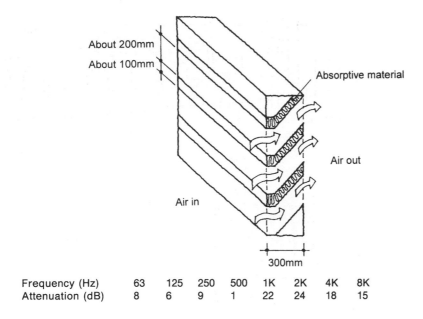

Frequency (Hz)	63	125	250	500	1K	2K	4K	8K
Attenuation (dB)	8	6	9	1	22	24	18	15

4.7 Cut-away view of acoustic attenuators and approximate attenuation performance.[20]

solution and gives attenuation data. (Note that 'bending' the air path makes it more difficult for sound to enter directly.)

Performance is worst at low frequencies. This is generally true for attenuators and is one reason why most of us live with a background low-frequency rumble. The attenuation can usually be improved by incorporating bends in the air paths, but this increases the resistance to air flow and so makes it less suitable for natural ventilation systems with their low driving forces; however, some simple mechanical ventilation added to the natural ventilation can be of assistance (Chapter 9).

Of course, in many situations noise levels will not be very high and so attenuation will be less necessary. It may also be possible to draw air from a quieter area around the building (and which may also have a higher air quality). In other situations, some background noise will be desirable. In university study bedrooms, for example, eliminating external noise (whether from traffic, the wind or passers-by) can be disconcerting and will tend to exacerbate the irritation due to noise from adjacent study bedrooms. In open plan offices, as noted in Chapter 2, some background noise is usually desirable because it helps mask individual conversations. Also, individuals often prefer to have some control over their environment and will choose a little more fresh air along with some more noise in preference to less noise but higher temperatures.

Each situation requires individual analysis. It does appear, however, that there is an argument for separating out the traditional functions of the window so that light, ventilation and noise can be dealt with more effectively. Too many 'solutions' are currently trying to do too much in too tight a space. Figure 4.8 shows a composite theoretical 'window' or, more accurately, wall which illustrates some options and future directions.

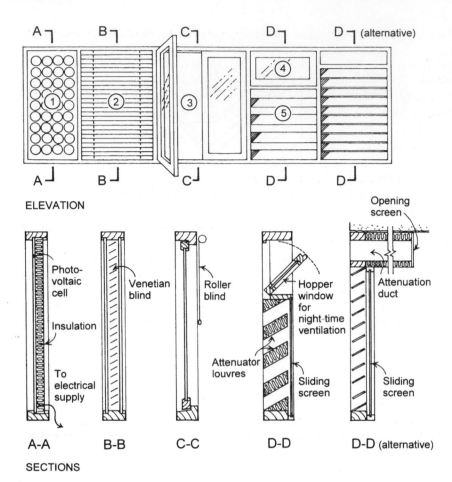

ELEVATION

SECTIONS

4.8 Composite 'window'.

Key
1. Future power section, e.g. photovoltaic cells (Chapter 7).
2. Fixed glazing with mid-pane blinds: allows variable daylighting level and control of solar gain.
3. Openable glazing (numerous types available – casement window shown) allows ventilation (but admits noise).
4. Night-time ventilation from manually openable hopper window; provides security and protection from rain.
5. Opaque section provides ventilation and acoustic attenuation. Light penetration can be minimized. Additional attenuation can be provided at soffit level if required, as shown in section 'D-D alternative'.

Obviously, possibilities and permutations are numerous. Where blackout facilities are required all blinds can be specified as such. If external solar control is desired it can easily be incorporated, and if light shelves (Chapter 8) or other such devices are needed they too can be added. As active noise controllers (systems which create equal and opposite pressure variations at the same frequencies) improve and become cost effective they may start to replace (or complement) the passive devices shown in the figure.

For more sophisticated (and costly) applications the functions shown can be motorized. Instead of a sliding screen, as shown in Section D–D, automatic dampers could be used and all opening windows could be motorized. These could be further linked to noise detectors or even odour detectors as these become commercially available in the future.

Window design, which is exceptionally important, is difficult and is treated in more detail in Chapter 9.

Having examined the main issues we can ask: What percentage of a wall should be glazed? Unfortunately, the answer is not straightforward. It depends on the building type (office or domestic), the building 'body', internal loads and so forth. In order to increase natural lighting in multi-storey offices, and thus reduce the energy consumption of artificial lighting, a large percentage of the wall needs to be glazed (Chapter 8). There are then at least two possibilities. The first is that adequate control at the building envelope plus natural ventilation pathways (with possible mechanical assistance) result in sufficient comfort. Alternatively, the space begins to overheat and the energy requirement for mechanical ventilation and, perhaps, air-conditioning, increases. Obviously, the first possibility is preferable – the principle to follow is designing to make the best use of available energy and incorporating controls to ensure that the advantages do not become nightmares. Unfortunately, there are no simple answers.

Second 'skins', buffer spaces and atria

It is also possible, of course, to put a second 'skin' or envelope around the first. This can help reduce heat loss while maintaining most of the benefits of solar gain directly into the building. It can also trap potentially useful solar heat between the skins. Acoustic attenuation is a further potential advantage. Buffer spaces like conservatories are localized second 'skins'.

Disadvantages of a second envelope include cost, possible interference with natural ventilation and creation of a large enclosed void connecting the building's floors through which smoke and fire can spread. Any proposed design must carefully analyse these issues.

Other less complicated buffer zones include draft lobbies and auxiliary spaces such as garages. In these the environment is uncontrollable or only loosely controlled but protection is afforded to the primary space. Atria are 'buffer' spaces that can vary widely in complexity. In their simplest form they are enclosed spaces that keep the rain out, allow light and solar radiation to enter and have no artificial heating; these simplest forms are also likely to have high level openings or extract systems to dispel heat and smoke. Complex atria tend to have installed heating systems and ventilation systems that interact with the spaces around the atria. They can also incorporate solar and light controls.

Atria have the potential environmental advantages of allowing single-sided and cross ventilation (Chapter 9) and passive solar gain and daylighting. However, the overall energy balance will depend on the specific design. More complex atria with artificial heating may be of limited benefit.

Control at the building envelope

Control is necessary because solar gain, temperature and wind speed vary so much. Traditionally, occupants have been able to influence their environment and comfort by simple, easy-to-use, robust means and were then able to see, and almost immediately experience, the results of their actions. An example is the Victorian hospital window with a tall sash and a top hopper window operated by a crank whose control was at nurse height. As we develop more sophisticated systems it is important to try to keep these principles in mind. For example, opening inlets and outlets for natural ventilation systems that cannot be seen by the occupants introduce an element of uncertainty that poses problems for the psychological perception of comfort.

Controls can, of course, be manual or automatic or some combination of both. One welcome development is more use of intelligent controls that allow occupants to override automatic systems for limited periods and then readjust according to conditions.

4.5 Internal layout

The internal layout joins the 'skin' to the 'body'. If there is only one space, or, in architectural slang, just one 'shed', the link is direct. In this case ventilation is straightforward, the noise is whatever comes through the skin and solar gains are relatively easy to deal with because they are entering a large volume.

As a space is subdivided, the situation alters. Partitions, particularly the full-height type, reduce the scope for natural ventilation, but ducted fresh air, probably with some mechanical assistance, can help overcome the problem. Opaque partitions interfere with views and reduce daylight penetration. However, partitions are likely to improve the acoustics by providing more privacy for conversations. Partitions and furniture also increase the admittance of a building - a figure of 1 W/K per m^2 of floor area has been estimated[21] – and this too can be an advantage.

Where loads are particularly high and daylighting requirements are low, as is often the case in lecture theatres – it could be advantageous to locate the space on the north side of the building, thus reducing solar gains.

Smoke and fire considerations will be mentioned in Chapter 9.

4.6 Form revisited

We can now consider form in greater detail and in particular look at two contrasting strategies of articulated versus compact forms.

An example exists in the UK where numerous government departments have been housed in collections of single-storey huts erected in the 1940s. With time these have been and continue to be replaced by multi-storey monoliths for economic reasons that include freeing land for more intensive development and lower building costs.

Environmentally, however, the arguments are less clear and somewhat contradictory. Compact shapes have low surface area/volume ratios and so are favoured where heat loss is a major issue – Inuits in their igloos know this. But there is less solar gain and daylighting available with the compact shapes and so, for example, energy consumption for artificial lighting will be higher. The significant potential for day-lighting from the roof of single-storey buildings is one of their great advantages. Taller buildings will be more exposed to wind infiltration and will require energy for lifts.

Natural ventilation is much easier with articulated forms, which provide more possibilities for single-sided and cross ventilation (Chapter 9). Views out and contact with a natural or landscaped environment are also favoured by articulation.

Once again, the optimum strategy will depend on the application. Schools and office buildings require higher levels of lighting and higher ventilation rates than homes and so there is more call for articulated shapes with greater surface areas and more glazing. In a sense, the 'skin' services the building. One wants as much 'skin' as can usefully be employed but no more, because more 'skin' also means greater heat loss. We shall return to these questions in Chapters 8 and 9.

4.7 Two (more) models

Butterflies (Figure 4.9) are lightweight with powerful wings of large area compared with their bodies. Their sense organs vary from eyes for vision to antennae for smell. Butterflies are quick to respond to their environment. Elephants, on the other hand, are not lightweight. They keep a wary eye on their surroundings and are loathe to forget. If their environment changes, they also change – but only after a period of time.

Butterfly-type buildings will have highly responsive skins with a great deal of glass (and other materials yet to be developed) and will react quickly to changes in solar radiation, light and temperature, by altering their properties. They will also have ventilation openings that vary according to constantly changing needs. Parts of their envelopes will capture energy and generate power or heat directly just as some butter-flies use the sun's radiant heat to warm themselves up so that they can get the full power of their muscles for flight.[22] Thermal mass will be incorporated but no more than necessary for vital building functions. Butterflies are 'high-tech'.

Elephant-like buildings have much more thermal mass. The building envelope is less critical because the mass compensates for the lack of a quick response. There are fewer openings and many of these may be manually controlled. Elephants are 'neo-vernacular'.

4.9 Butterflies and elephants.

Of course, most buildings in the next few decades will be something in between as specific requirements for noise attenuation or heat disposal or increased views guide us to real designs.

A final word: as might be expected, the analogy does not support rigorous analysis. Butterflies are poikilothermic, i.e. their temperatures vary with those of their surroundings. Elephants are homeothermic and maintain a constant body temperature by internal means. Buildings with heating systems that work are homeothermic; those with systems designed by inexperienced engineers are poikilothermic.

Guidelines

1. Orientate the building to the south if possible.
2. Incorporate the right amount of thermal mass and high admittance surfaces into the building.
3. Increase the floor-to-ceiling heights in heavyweight buildings. Remember, the more height, the more light will enter.
4. Use glazing to allow solar gains and daylight but control at the building envelope to avoid overheating and glare.
5. Incorporate a suitable degree of air tightness.
6. Specify windows and doors that are suitable for the degree of exposure and are detailed to reduce infiltration losses through them.
7. Insulate well to reduce heat loss.
8. Consider shutters or curtains to reduce night-time heat losses.
9. Decide if noise is a problem and, if so, how it will be attenuated.
10. Choose a compact or articulated form, or something intermediate according to suitability.
11. Use simple buffer spaces to reduce heat loss.

References

1. Milbank, N.O. and Harrington-Lynn, J. (1974) Thermal response and the admittance procedure. Current Paper 61/74. BRE, Garston.
2. Loudon, A.G. (1968) Summertime temperatures in buildings without air-conditioning. Current Paper 47/48. Building Research Station, Garston.
3. Anon. (1999) *CIBSE Guide A3: Thermal Properties of Building Structures*, CIBSE, London.
4. See reference 1, p. 40.
5. See reference 3.
6. See reference 1, pp. 47–50.
7. Anon. (1994) Minimising/avoidance of air-conditioning. Final Summary Report. Revision A. BRE Project EMC 32/91, Max Fordham & Partners, London.
8. Bordass, B., Entwisle, M. and Willis, S. (1994) Naturally ventilated and mixed-mode office buildings: opportunities and pitfalls. *CIBSE National Conference Proceedings, Brighton*.
9. Data from Anon. (1988) *Solar*, Monsanto, St Louis.
10. Anon. (1992) *Pilkington K Glass and Kappafloat*, Pilkington, St Helens.
11. Anon. (2004) *The glass range for architects and specifiers*, Pilkington, St Helens.
12. Pilkington Glass (2004) Private communication.

13. Anon. (1993) Double glazing for heat and sound insulation. BRE Digest 379. BRE, Garston.
14. www.swedishconstructionproducts.co.uk, 2 March 2005.
15. Anon. (1979) *How windows save energy*, Pilkington, St Helens.
16. Banham, R. (1969) *The Architecture of the Well-Tempered Environment*, The Architectural Press, London.
17. Anon. (1987) *Window Design: GIBSE Application Manual*, CIBSE, London.
18. Anon. (1981) Cited in CIBSE Building Energy Code, Part 2, CIBSE, London.
19. Anon. (1993) Double glazing for heat and sound insulation. BRE Digest 379. BRE, Garston.
20. Data from Airstream, Wokington, Berkshire.
21. M. Entwisle, Max Fordham & Partners (1994) Private communication.
22. Wigglesworth, V.B. (1964) *The Life of Insects*, The New American Library, New York.

Further reading

Daniels, K. (1997) *The Technology of Ecological Building*, Birkhäuser, Basel.
Edwards, B. and Tuppent, D. (2001) *Sustainable Housing Principles and Practice*, Spon, London,
Hawkes, D., McDonald, J. and Steemers, K. (2002) *The Selective Environment*, Spon London.
Lewis, S. (2005) *Front to Back–A Design Agenda for Urban Housing*, Architectural Press, Oxford.
Lloyd-Jones, D. (1998) *Architecture and the Environment*, Lawrence King, London.
Porteous, C. (2002) *The New Eco-Architecture*, Spon, London.
Roche, L. (1997) Smart glass. *Building Services Journal*, **19**(8), 27–9.
Smith, P. (2003) *Sustainability at the Cutting Edge*, Architectural Press, Oxford.

CHAPTER 5

Site planning

5.1 Introduction

Around 30 B.C. Vitruvius wrote of the need to choose the most temperate regions of climate, since we have to 'seek healthiness in laying out the walls of the city' and went on to say that 'the divisions of the sites . . . the broad streets and the alleys . . . will be rightly laid out if the winds are carefully shut out from the alleys. For if the winds are cold they are unpleasant. . .'[1]

As Vitruvius knew, the site will have a marked effect on the functioning of the building and the building, in turn, will affect the site. Issues of particular concern are:

- site selection, microclimate, landscaping and biodiversity
- sunlight, solar gain and photovoltaics
- daylight and views
- wind
- noise
- air quality.

5.2 Site selection, microclimate, landscaping and biodiversity

On a broad scale this book principally addresses issues for what has been termed the mid-European coastal climate as shown in Figure 5.1.

The European climate zones, which have fairly fuzzy boundaries, have been described as follows.[2]

1. Cold winters with low solar radiation and short days; mild summers.
2. Cool winters with low solar radiation; mild summers.
3. Cold winters with high radiation and longer days; hot summers.
4. Mild winters with high radiation and long days; hot summers.

One could add comments about the wind because of its importance for natural ventilation: the mid European coastal zone is characterized by a strong wind regime – at Heathrow airport in London, for example, a wind speed of 4 m/s is exceeded more than 50% of the time[3] (Figure A.3).

If we now consider sites and assume that a building site is required, consideration should be given to reusing existing ones if free, or to selecting locations that are least likely to be damaged environmentally if built upon. A checklist developed by the Building Research Establishment (BRE) for office buildings contains an examination

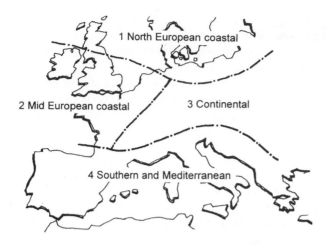

5.1 European climate zones.[4]

of whether the site includes ecologically valuable features such as mature vegetation, ponds or streams and natural meadows and, if so, effectively attempts to direct developers towards other areas of lower ecological value.[5]

The next step is to try to create a combination of building and site which marries the aesthetic and physical environments. The elements of the physical environment include the site layout, the form of the building (in both plan and section), the materials used externally for the building and the landscaping (both hard and soft). Together, these factors will create a microclimate.

Traditionally, the primary concern of the microclimate in the UK has been to mitigate the effect of the 'cold, wind and wet of the relatively long cool season'.[6] However, we are in an exceptional period of transition to more sustainable buildings and groups of buildings. Keys to this are maximizing the solar potential of the site and using wind energy when appropriate. This is a shift from a more traditional, 'defensive' strategy to a more responsive, 'active' one. There will be conflicts to resolve in this process. Designing for higher wind speeds to power turbines or, say, for natural ventilation (see Chapter 9) in the summer could mean ventilation heat losses in the winter are slightly greater. Other potential conflicts to consider are:

– landscaping reduces wind speeds which could be advantageous or disadvantageous; it can also cause a loss of solar energy and natural light if not used carefully
– the use of dark surfaces near buildings to absorb solar radiation or light surfaces to reflect light into them, and
– water features that detrimentally dampen conditions in winter but are advantageous in helping produce a cool microclimate in summer.

Beneficial microclimates in the heating season are those that create warmer, dryer conditions and this can be done in a number of ways. The first is by taking advantage of solar gain (discussed principally in section 5.3), reducing the wind speed (discussed in section 5.5) and lessening the effect of rain. Rain can be dealt with by effective surface water drainage systems. These should be designed in accordance with the principles of what has become known as Sustainable Urban Drainage. The key idea is to manage surface water run-off in more natural ways to reduce the possibilities of flooding and pollution. Concepts and techniques include the use of permeable surfaces, filter strips, swales and ponds.

Landscaping has an important role to play in drainage and also in encouraging biodiversity, i.e. the complex variety of habitats and organisms that both surrounds us and includes us. In the next few decades we can expect to see a more 'biological' age replace the age of the machine and closer integration of plants, animals and buildings.

Building form can also play an important role in controlling rain. For example, at Calthorpe Park School in Hampshire, 1200 mm deep roof overhangs were used both as sunscreens and shields to keep rain off the building fabric, thus protecting it and improving its insulating value.[7]

5.3 Sunlight and solar gain

Buildings can, of course, be located completely or partially underground, as shown in Figure 5.2; as well as more traditionally above ground. Careful attention needs to be given to drainage, daylighting and ventilation of underground buildings, but fabric heat losses will normally be reduced because soil temperatures during the heating season are higher than air temperatures. Orientating partially underground buildings to the south will allow passive solar gain to contribute to the (reduced) space-heating requirements.

5.2 Ecology House at Stow, Massachusetts.[8]

A variation on this theme is the single-storey building with extensive landscaped roof gardens and spacious courtyards that allow solar gain into south-facing spaces even in the winter. An example of this is discussed in Chapter 11.

Most buildings have been – and will continue to be – built above ground and for these one question is how to make the best possible use of solar gain in order to reduce energy consumption. Some use of solar gain is already made – it has been estimated that the Sun provides about 14% of the space-heating demands on average in UK homes.[9] Note that this is not all from direct solar radiation, but that an important contribution is made by diffuse solar radiation. It is easier in assessment techniques, however, to concentrate on direct radiation, and that is the approach followed below.

The Sun's position, of course, varies throughout the year – in London at a latitude of 51.5 °N the solar elevation on 21 December is 15° at noon and this rises to about 62° at noon on 21 June.

Figure 5.3 shows the minimum north/south spacings required to give solar access at noon. Because before noon and after noon the solar altitude will be less, increasing these spacings will increase the number of hours of *solar access*. The direct radiation referred to in the figure is that portion of the solar radiation which comes directly through the atmosphere; sky diffuse radiation is that portion which is scattered back to Earth from the atmosphere.

Access to the Sun has both psychological and physiological effects that have always been appreciated. Figure 5.4 shows the magnificent sixteenth-century refectory of Fontevraud Abbey in France where solar gain through the large windows on the left of the figure was used to cheer the souls of the nuns who ate there.

A number of computer programmes for analysing solar access[10] now complement early manual design tools.[11]

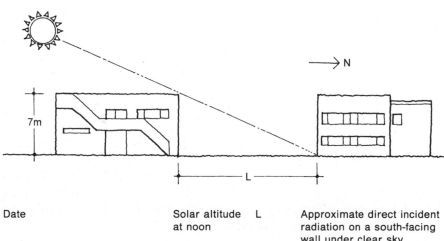

Date	Solar altitude at noon	L	Approximate direct incident radiation on a south-facing wall under clear sky conditions at noon
	(deg.)	(m)	(W/m^2)
21 December	15	25.9	500
21 March and 22 September	38	9.0	630

5.3 Spacings to achieve solar access at noon and radiation data for London.

5.4 Refectory at Fontevraud Abbey.

5.5 Housing at Angers, France.[12]

Work over the past 30 years has tended to concentrate on how to arrange large groups of houses to optimize use of the Sun's energy, and a typical layout is shown in Figure 5.5.

In such schemes a starting point is to get the road pattern right (roughly east–west); correct spacing should be dealt with at the same time. Common guidelines are to space houses in England more than twice their height apart and to orientate the long axis of the house within 45° of south.[13] If possible it is even better to orientate the

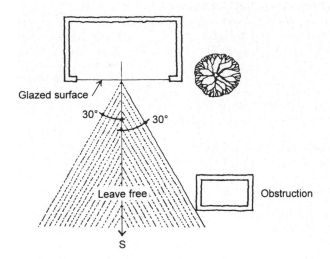

5.6 Orientation for passive solar gains in winter (based on reference 14).

house due south and to keep the sector 30° on each side free of obstructions, as shown in Figure 5.6.

For larger, more communal projects, the location of open spaces, gardens, court-yards, garages and stores offers scope for facilitating solar access.

Trees and other vegetation have an important role to play in site layouts because of their amenity value and effect of tempering the wind (section 5.5). They can also provide some control of summer-time solar gain to avoid excessive temperatures at a cost of a winter-time loss of passive solar gain and a year-round loss of light (such trees effectively function as permanent, albeit seasonally variable, fixed external shades). Figure 5.7 shows a typical situation.

Any trees selected should, of course, be suited ecologically to the site. The designer can then consider, for deciduous trees, how long they are in leaf and how transparent they are to solar radiation, both in leaf and bare. Table 5.1 provides a selection of such data.

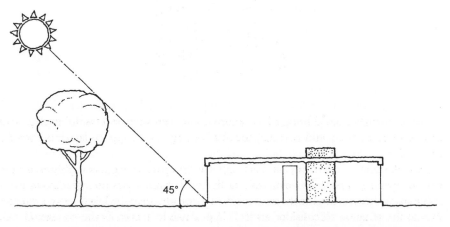

5.7 Effect of trees on solar access.

Table 5.1 Characteristics of common deciduous trees in the UK [15]

Botanical name	Common name	Period of full leaf	Transparency (% radiation passing)	
			Full leaf	Bare branch
Acer pseudoplatanus	Sycamore	May/August	25	65
Aesculus hippocastanum	Horse chestnut	Mid April/August	10	60
Betula pendula	European birch	May/August	20	60
Quercus roba	English oak	Mid May/mid October	20	70
Ulmus	Elm		15	65

Notes
1. Data is based on averages. Wide individual variations exist and so caution should be exercised.
2. Measurements are usually based on light but can be used for solar radiation also.

Thus, if we choose an elm for our tree in Figure 5.7 and locate it so that it will block out 85% of the Sun's radiation when in leaf in the summer (i.e. 15% transparency), it will still block out 35% during the winter. Depending on the design, this can be a strong argument for less permanent solar shades.

Nonetheless, there tends to be a quite reasonable compromise between the amenity value of the trees and their functional role as windbreaks and solar screens. One approach is to locate deciduous trees to the south of southerly orientated buildings and to site lower evergreens to the north as a windbreak. The lower evergreens can, of course, also be used around the site for privacy and to the south as a windbreak (section 5.5).

5.4 Daylight and views

Daylighting of a space through a window is a function of the amount of sky the building can 'see' and, to a lesser extent, reflection from the surrounding surfaces. Assessing daylight access is somewhat similar to assessing solar access but differs in applying to surfaces at all orientations. Existing assessment techniques consider daylight availability, the effects of external obstructions and the reflectivity of external surfaces (Chapter 8 and Appendix B).

Daylighting guidelines exist for new developments. The BRE suggests that, as a first step, a check is made to see if there are obstructions within 25° of a reference line,[16] as shown in Figure 5.8.

If obstructions are at less than a 25° angle, the BRE advises that there will be potential for good daylighting in the interior, and an obstructing building that is 'too tall' but narrow may still permit good daylighting. Often site constraints, however, will not allow this criterion to be met. At the De Montfort Queens Building (Chapter 12), the spacing between the electrical laboratory wings was narrow (Figure 5.9) but by using white high-density panels as the external cladding the architect, nonetheless, achieved a light feeling in the courtyard and reasonable light levels in the interiors. It may be of value to consider an 'external' daylight factor in these cases. Point P

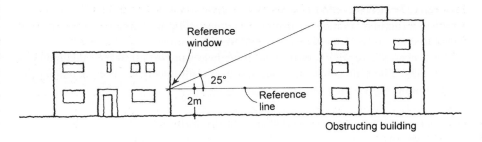

5.8 Angular criterion for spacing of building.

receives about 40% of the light incident above the building at point *A* in overcast sky conditions.

In even more constrained situations, as in many urban developments, the primary consideration is, in fact, ensuring that any new development does not affect its neighbours' 'right to light', and planning consent may depend on a successful solution to this problem.

We have seen how vegetation can have an adverse effect on solar gain in the winter, and since daylight is one part of solar radiation, it will, of course, be similarly reduced.

Just as a 'right to light' exists effectively for buildings, a right to a view ought to exist for people. Picasso used to say that he liked a view but preferred to sit with his back to it; most of us, however, would prefer some contact with the outside, whether this is to see changing sky conditions or panoramic scenes of cities. In numerous projects, from factories to restaurants to, more conventionally, schools, we have found that views out have been a major ingredient in the building's success.

5.5 Wind

A striking manifestation of wind forces on buildings is the flying buttresses of medieval cathedrals. As the cathedrals grew taller the architects and engineers found that

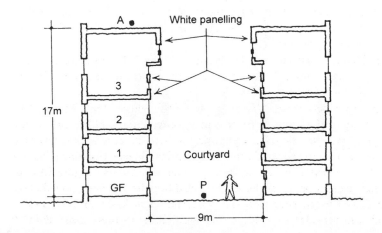

5.9 De Montfort University's Queens Building: electrical laboratory spacing (simplified).

wind forces (which are proportional to the square of the velocity of the wind) required a radical structural solution to maintain stability. The first flying buttresses were introduced at Notre-Dame de Paris in the twelfth century.[17]

Our concerns here are less critical but nonetheless important to the functioning of the building. The main aim in the heating season traditionally has been to temper the winds around the site in order to:

- reduce the infiltration of external air into the building;
- increase the surface resistance of elements such as glazing, thus improving its thermal properties (eg the U-Value for double glazing at an 'exposed' site is about 3.2 W/m^2 K and at a 'sheltered' site about 2.8 W/m^2 K);
- reduce the wetting of the fabric by wind-driven rain, thus helping to maintain its insulation properties (both as a resistance to conductive heat transfer and by keeping it dry, thus stopping evaporative heat loss).

Obviously, reducing the wind speed on a site will also make it a more environmentally comfortable and enjoyable space for those who use it.

It is important that if the building is to be naturally ventilated any measures taken to improve the winter condition do not worsen the summer one. In principle, this should not prove too difficult; the main consideration will be ensuring that the paths to the air inlets are relatively free. Air outlets will normally be higher (Chapter 9) and should pose less of a problem. Wind is also useful in carrying away heat and pollutants from a site, and enough movement must be retained to ensure this.

Designers have one main way of tempering the winds and that is the use of windbreaks, which are likely to be of vegetation but can also include, for example, fences and other buildings. In laying out a site and incorporating windbreaks or shelterbelts care has to be taken to ensure that vegetation, in particular, does not significantly reduce passive solar gain. Figure 5.10 shows an idealized shelterbelt for protection from westerly winds.

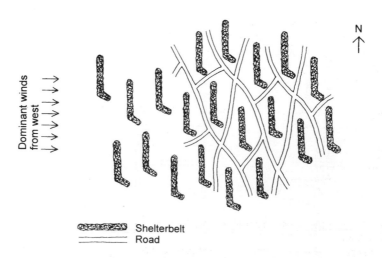

5.10 Idealized shelterbelt layout for protection from westerly winds.[18]

Dominant winds from west

N

Shelterbelt
Road

In the UK, the dominant winds are generally from the southwest and northwest but they vary significantly with place and time.[20] Designers should not rely on the wind being from a particular direction; however, they should try to determine the dominant wind direction in the summer and winter and draw the likely air movements on their site plans to help them visualize the flow patterns. Some data on wind is given in Appendix A.

The height of the shelterbelt and its permeability determine the area protected. Figure 5.11 shows an idealization of the effect of a windbreak.

Thus, if a shelterbelt is 10 m high, then up to a distance of about 120 m, the wind velocity will be below 50% of the reference wind velocity, which in Figure 5.11 is 9.2 m/s.

A number of rules of thumb exist. For example, a guide for energy efficiency in new housing suggests to reduce wind and to allow solar access to the site,[21] a shelterbelt should be located at a distance of three to four times its height from the homes to be protected. As always there is a balance to be struck among many factors. If, for example, the shelterbelts are far apart (to allow for solar gain), the increased wind speed may negate the effect of the increased solar input to the building. The goal is to optimize the overall performance of the microclimate of the site and buildings, and this remains more of an art than a science.

Designers can also influence the wind speed around a building through its form. The objective is to approximate forms that present the least resistance to the passage of the wind around them, thus reducing the disturbance to the wind pattern near the ground. It has been suggested that for normal, rectilinear buildings this implies a shape that is as near as practical to a pyramid.[22] Methods vary, from using hipped roofs rather than gable roofs for houses to stepping back the façades of multi-storey buildings. It must be said, however, that a pyramid has a greater surface area/volume ratio than a cube or a compact parallelopiped and so, inevitably, these techniques tend to increase surface area and raise the heat loss. They also imply a reduction in area suitable for photovoltaic panels and solar collectors. Where the

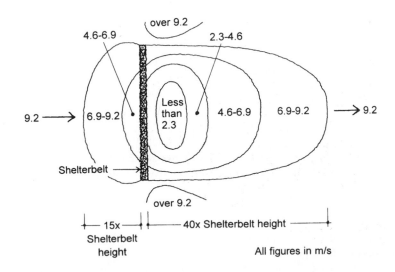

5.11 Reduction in wind due to a good shelterbelt.[19]

balance lies will depend, in the by now familiar way, on the site and on a variety of other factors.

Groupings of buildings can also be made less sensitive to the wind by using irregular patterns (Chicago became 'The Windy City' in part because of its regular street pattern, facilitating winds off Lake Michigan), keeping heights of buildings fairly uniform and creating courtyards. These and other techniques are described in references 23 and 24.

In addition to all of these considerations, the possibility of using wind energy on a site needs to be reviewed. Although wind mills have been common for centuries in the countryside and city suburbs (think of Paris' Moulin Rouge) the use of wind energy for electricity is a relatively new phenomenon that reflects our need to replace fossil fuels with renewable energy sources. References 25 and 26 describe some of the possibilities and Chapters 15 and 17 describe some applications.

5.6 Noise

Careful arrangement of buildings and the use of shelterbelts can also improve the acoustic aspects of a site. Attenuation can vary from 1.5 to 30 dB per 100 m of shelterbelt, depending on the type of vegetation in the shelterbelt.[27]

5.7 Air quality

The air quality of a site can be improved by ensuring that winds can cleanse it and through the use of vegetation.

In photosynthesis plants absorb carbon dioxide and produce oxygen, and through transpiration they absorb water at the roots and release it into the air, principally at the leaves. Plants can also cleanse or filter the air when dust and pollutants adhere to their dry twigs or leaves (which are eventually washed by rain and impurities are deposited on the ground). Thus, highly planted zones will have a higher oxygen content, higher relative humidity and fewer pollutants and are likely to provide the right type of area from which to draw the supply air for a natural ventilation system.

Guidelines

1. Select a suitable site.
2. By siting of the building and the use of landscaping, develop a favourable micro-climate with a suitable temperature, wind and relative humidity regime.
3. Encourage biodiversity.
4. Orientate and space buildings to make use of passive solar gain and daylight.
5. Provide occupants with views out.
6. Strike the right balance in tempering the wind where necessary and using it as a source of energy where possible.

7. Improve the noise climate on the site through grouping of buildings and the use of vegetation.
8. Remember that vegetation improves air quality.

References

1. Vitruvius. *The Ten Books on Architecture*, Book 1 (transl. F. Granger, 1931), Heinemann, London.
2. Baker, N. *et al.* (*ca.* 1992) *The LT Method: Version 1.2*, Commission of the European Communities.
3. Anon. (1999) *CIBSE Guide A2: External Design Data*, CIBSE, London.
4. See reference 2.
5. Anon. (2004) BREEAM Offices 2004. BRE, Garston.
6. Anon. (1990) Climate and site development. Part 2: Influence of microclimate. BRE Digest 350. BRE, Garston.
7. Davies, C. (1985) Hants improvements. *Architectural Review*, **188**(1056), 22/2–29/2.
8. J.E. Barnard, Jr (1982) Private communication.
9. Anon. (1993) Low energy design for housing associations. *BRECSU Good Practice Guide* No. 79. BRE, Garston.
10. Anon. (1992) Lighting for buildings. BS 8206: 1992. British Standards Institution, London.
11. See, for example, TAS by EDSL Ltd, and Ecotect by Square One Research Ltd, Cardiff University.
12. Dodd, J. (1989) Greenscape 2: Climate and form. *Architects' Journal*, **16**(189), 81–5.
13. See reference 9, p. 10.
14. Littlefair, P.J. (1992) Site layout for sunlight and solar gain. BRE Information Paper 4/92. BRE, Garston.
15. Anon. (1990) Climate and site development. Part 3: Improving microclimate through design. BRE Digest 350. BRE, Garston.
16. Littlefair, P.J. (1992) Site layout planning for daylight. BRE Information Paper 5/92. BRE, Garston.
17. Mark, R. (1990) *Light, Wind and Structure*, MIT Press, Cambridge.
18. See reference 15.
19. Anon. (1964) *The Farmer's Weather*, Ministry of Agriculture, Fisheries and Food, Bulletin No. 165. HMSO, London.
20. See reference 3.
21. Anon. (1993) Energy efficiency in new housing. *BRECSU Good Practice Guide* No. 79. BRE, Garston.
22. See reference 15.
23. See reference 15.
24. Dodd, J. (1989) Greenscape and tempering cold winds. *Architects' Journal*, **189**(18), 61–5.
25. Study by Ecolys and Delfi University of Technology, cited in Timmers, G. (2001), Wind Energy Comes to Town: Small Wind Turbines in the Urban Environment. *Renewable Energy World*, May–June, pp. 113–119.
26. Thomas, R. (2002) *Sustainable Urban Design*, Spon Press, London.
27. Johnston, J. and Newton, J. (1993) *Building Green*, London Ecology Unit, London.

Further reading

Anon. (2002) Sustainable urban drainage systems – design manual for England and Wales. Construction Industry Research and Information Association, London.

Dodd, J. (1993) Landscaping to save energy. *Architects' Journal*, **198**(1), 42–5.

Kudadia, V., Pike, J. and White, M. (1997) Air pollution and natural ventilation in an urban building. A case study. *Proceedings of the 1997 CIBSE National Conference*. CIBSE, London.

Littlefair, P. (2001) Solar energy in urban areas. BRE Information Paper. BRE, Garston.

McHarg, I. (1971) *Design with Nature*, Doubleday, New York.

Materials and construction CHAPTER 6

6.1 Introduction

'The subject of Material is clearly the foundation of Architecture', said William Morris in 1892[1] and now, over a century later, with a far wider range of materials at the designer's disposal and more awareness of the environmental impact of materials, the statement has added significance. Materials affect structure, form, aesthetics, cost, method of construction and internal and external environments. This chapter examines basic criteria for their selection and provides data on those commonly used in buildings. It includes brief discussions of some construction issues and environmental assessment techniques.

6.2 Selection of materials

We should ask what criteria should be used when selecting materials before examining any specific ones. Relevant considerations (which complement the usual ones such as fitness for the purpose, cost, mechanical resistance, stability and safety) include impact on the natural environment and impact on health, with the two often being related. The impact on the natural environment includes ecological degradation due to extraction of raw materials, pollution from manufacturing processes, transportation effects, energy inputs into materials which affects CO_2 production and the use of refrigerants. Health issues range from how materials are extracted to the effects on the manufacturing workers producing the materials and to the internal environment that results from the materials selected. The major topics are discussed below but it should be remembered that the entire field is in a state of flux, and as more is learned about materials and the environment, conclusions will change. To cite but one example: in the 1970s the Cambridge University Autarkic House project developed a home intended to minimize energy use and supply the much reduced demand with solar and wind energy.[2] The intention was to use 700 mm of polyurethane for the thermal store insulation, and only later was it realized that the CFCs used in the manufacture of polyurethane were a major environmental hazard.

6.3 Environmental aspects of materials

The subject is, of course, enormous, and below we touch on only a number of key issues. The references and further reading suggestions at the end of the chapter provide additional information.

Sustainable development has been defined, none too precisely, as 'development that meets the needs of the present without compromising the ability of future generations to meet their own needs'.[3] How much of which materials should we be using? is an unanswered question. Should we prohibit the use of rare materials or should they be acceptable when 'absolutely necessary'?

An important question is: Where do the materials come from? For building materials areas of concern include land-take for aggregates (crushed rock, sand and gravel). Few alternatives exist but some recycling is possible. Marine extraction of aggregates is eschewed by many as unnecessarily environmentally damaging.

Deforestation is another key issue but not a new one. England's forests have been reduced from about 1.8 million hectares in 1500[4] to about 1.0 million at present. Early uses of the timber included shipbuilding and fuel for iron-smelting and building. At present a great deal of work is being put into developing sustainable woodlands and accreditation systems for them. Designers should ensure that all wood is accredited by either the FSC (Forestry Stewardship Council) or a suitable alternative organization.

Manufacturing and processing form another broad area of concern. Timber preservation, for example, is often essential for longevity but the chemicals used in the process need careful selection and handling. Concern about disposal affects many products, including plastics. PVC (polyvinyl chloride), for example, represents about 25% of total worldwide plastics production and is widely used in buildings for sheathing electric cables and for drains, cladding, floor coverings and window frames. Environmental concern has focused on recycling and burning of chlorides and release of pollutants. The situation needs to be kept under review.

6.4 Refrigerants

This section is called refrigerants but in fact covers a somewhat wider scope.

An important environmental issue is the depletion of the ozone layer mentioned in Chapter 2. This has been caused principally by man's use of a number of refrigerants and halons (halogenated hydrocarbons with bromine).

Terminology in this area is important but unfortunately somewhat complex. CFC stands for chlorofluorocarbon and refers to an organic molecule with chlorine and fluorine atoms. A measure of the damage caused to the ozone layer is a substance's ozone depletion potential (ODP). The ODP of the CFC known as refrigerant R11 is defined as 1.0 and other refrigerants are referred to it – the closer the ODP is to zero, the better the refrigerant is environmentally. In addition to their harmful effect on the ozone layer, refrigerants also contribute to the greenhouse effect, or global warming, as we saw in Figure 2.5. Table 6.1 gives data for a number of refrigerants. As can be seen from the table, most of these substances are much more harmful greenhouse gases than CO_2, which explains why, although the volume of such gases produced is relatively small, they account for about 10% of global warming.

The Montreal Protocol referred to in Table 6.1 is the 1987 multi-national agreement on reduced refrigerant and halon emissions into the atmosphere.

HCFCs are hydrochlorofluorocarbons. They contain chlorine but have lower atmospheric lifetimes than CFCs and are less damaging to the ozone layer, as indicated by their lower ODPs.

Table 6.1 Characteristics of refrigerants[5, 6, 7, 8]

Sub-stance	Type	Formula	Montreal Protocol	Ozone depletion potential $(R11=1)$[a]	Global warming potential[a] $(CO_2=1)$	Flamm-ability
R11	CFC	CCl_3F	Y	1	1600	No
R12	CFC	CCl_2F_2	Y	1	4500	No
R22	HCFC	$CHClF_2$	(N)	0.05	510	No
R113	CFC	CCl_2FCClF_2	Y	0.8	2100	No
R114	CFC	$CClF_2CClF_2$	Y	1.0	5500	No
R115	CFC	$CClF_2CF_3$	Y	0.6	7400	No
R123	HCFC	$CHCl_2CF_3$	(N)	0.02	29	No
R124	HCFC	$CHClFCF_3$	(N)	0.02	150	No
R125	HFC	CHF_2CF_3	N	0	860	No
R134a	HFC	CF_3CH_2F	N	0	420	No
R141b	HCFC	CH_3CCl_2F	N	0.08	150	Slight
R142b	HCFC	CH_3CClF_2	(N)	0.06	540	Slight
R152a	HFC	CH_3CHF_2	N	0	47	Moderate
R407c	HFC-407			0	1950[c]	
R152a	HFC-152a			0	190[c]	
R404A	HFC blend			0	4540[c]	
R410A	HFC blend			0	2340[c]	
R413A	HFC blend			0	2180[c]	
R417A	HFC blend			0	1950[d]	
R500		R12/R152a	Y[b]	0.74	3333	No
R502		R22/R115	Y[b]	0.33	4038	No

[a] Global warming and ozone depletion potentials are per unit mass, and values are current best available estimates which may be subject to revision. Global-warming potentials relate to the long-term (500-year) warming potential. (N) in the Montreal Protocol column means that the substance is an HCFC and is expected to be phased out between 2020 and 2040 or earlier as alternatives are developed.
[b] R500 and R502 are implicitly included in the Montreal Protocol because they contain the restricted refrigerants R12 and R115.
[c] Values vary with the reference; data from reference 9.
[d] Values vary with the reference; data from reference 10.

Nonetheless, following the phasing out of CFCs, we are now in a period of phasing out of HCFCs.

HFCs are hydrofluorocarbons. They contain no chlorine and have a negligible effect on the ozone layer (the ozone depletion potential of HFCs is estimated to be a thousandth of that of R11[5]) but they do contribute to global warming.

Because of this, a number of countries prefer that they be banned and we are likely to see their gradual phasing out also in the future.

These trends are promoting interest in a wide range of more natural refrigerants such as ammonia, hydrocarbons (HCs) and carbon dioxide. These substances, some of which have been used as far back as the nineteenth century, have a promising future.

Refrigerants, as we saw at the beginning of the chapter, have been used as blowing agents for plastic insulation materials. CFCs have been phased out and HCFCs are being phased out. Attention is now focused on HFCs (with their global warming potential). Alternative blowing agents which are more appropriate include hydrocarbons and carbon dioxide.

Generally, designers should consider how their buildings will be constructed and deconstructed. They should also allow for upgrading of the building during its lifetime.

6.5 Materials and health

Materials extraction and product manufacture are critical health and safety points. It is worthwhile noting that our environmental and health problems are not, sadly, novel. The charming churches of Norfolk are often built in knapped flint (i.e. fragments derived from nodules of almost pure silica). The knappers often worked in conditions of poor ventilation in an atmosphere of fine dust which caused silicosis and the premature death of many of them.[11]

Asbestos has been known to be a major hazard to health for some time and can no longer be specified for use in buildings. However, its disposal is a common problem that needs careful consideration when existing buildings are refurbished or demolished. This reminds us of the need to consider the full life cycle of any material or energy source.

Materials of high radioactivity should obviously be avoided because of the health hazard, but there are many other materials for which the danger is not necessarily as evident. These vary from products which release formaldehyde, those manufactured with or incorporating certain solvents, timber products treated with hazardous chemicals (for example, the insecticide HCH known as lindane) and composite materials incorporating certain resins. Each needs to be judged on its own merits.

Some paints traditionally have incorporated toxic metals such as cadmium (cadmium yellow was a favourite of the Impressionist painter Monet in his later works) and lead. Lead-free paints are now mandatory except in listed buildings.

Health in the workplace is a major environmental issue which reflects the enormous amount of time we spend in these relatively sealed areas. Indoor air quality (mentioned briefly in Chapter 2) is part of this. Another area of concern that needs to be monitored is the effect on health of electromagnetic fields due to electrical distribution systems and electrical equipment, including such mundane devices as hairdryers.[12]

6.6 Materials and energy

On 3 February 1695 at Versailles inside the Hall of Mirrors it is said that the temperature dropped to the point where wine and water froze in the glasses. It was an exceptionally cold year, but even in more clement times the heating system – consisting of two open fireplaces – consumed and furnished only relatively small amounts of energy. The energy that had gone into the splendid stone and decorations of the Hall, however, was significant. Both the running energy and initial energy were derived mainly from renewable sources, in particular wood, water and wind power; coal was available and had, for example, been used at least since the twelfth century for lime production,[13] but its cost limited its employment.

By contrast, energy inputs for running buildings now tend to be much greater than the initial energy inputs. Initial energy, or, more precisely, embodied energy, has been defined as the energy used to (a) win raw materials, (b) convert them to construction materials, products or components, (c) transport the raw materials, intermediate and final products; and (d) build them into structures.[14] The figures do not include maintenance, reuse or final disposal. Determination of the embodied energy is a field fraught with uncertainty for a number of reasons, including the difficulty of standardising data and incomplete knowledge of the fuel mix used in production. The field is also rife with debate as manufacturers stake rival claims to lower embodied energy and, thus, lower 'embodied' CO_2 production. Furthermore, it is an area that changes as manufacturing processes evolve. As a consequence, emphasis (and availability of data) has shifted to environmental profiles. Some data for embodied energy is provided but as it is dated it is for indicative purposes only.

In the UK, approximately 5–6% of the total energy consumption is embodied in construction materials[15] compared with about 50% used in buildings for space heating and cooking, water heating, lighting and power (Chapter 3).

For new office buildings as a whole, the embodied energy ranges from 3.5 to 7.5 GJ/m^2 of floor area whereas energy in use amounts to between 0.5 and 2.2 GJ/m^2 yr; typically, the initial embodied energy of an office is equivalent to about five years of energy in use, or about 7% of the total energy used over the lifetime of the building.[16] Obviously, as buildings become more energy efficient in use, the embodied energy will become relatively more important and, similarly, the relative energy involved in demolition and the importance of recycling materials will increase. At present, however, the greatest energy savings are to be obtained by reducing energy consumption in use.

In this field of embodied energy it is useful to try to find a position from which to take an overall view. Table 6.2 gives broad worldwide and UK comparative energy requirements for major building materials.

The table must be used cautiously. As is evident, there are major variations, although the broad classification of energy bands seems about right. Generally, energy inputs will depend on a country's fuel mix, and the source of a country's materials will affect the figures. For example, timber varies depending on whether or not it is grown locally. Most softwood in the UK is imported and so the embodied energy includes a significant transportation component. Energy inputs into metals such as copper and aluminium can vary widely according to whether the source is from the ore or from recycled material.

If we now look at the main structural materials used in building, say, a typical detached house, we find that the bulk of the building relies on quite a small number of materials and that the embodied energy requirement of very-high-energy and high-energy materials is not a great percentage of the total. Figure 6.1 shows typical embodied energy inputs.

Obviously, the materials and the relative contributions of the various components will vary with building type and particular buildings. Reference 20 gives an analysis of energy inputs into major building materials (glazing, finishes and services are all excluded) for local authority housing types, varying from a two-storey house to a nine-storey block of flats. In the former, steel reinforcement accounts for 1% of the total of components analysed and in the latter it is 40%!

Table 6.2 Broad comparative energy requirements of building materials

Material	Primary energy requirement (GJ/tonne)		
	Worldwide[17]	UK[a18]	UK[19]
Very-high-energy			
Aluminium	200–500		97
Plastics	50–100		162
Copper	100+		54
Stainless steel	100+	75[a]	
High-energy			
Steel	30–60	50	48
Lead, zinc	25+		
Glass	12–25		33
Cement	5–8		8
Plasterboard	8–10		3
Medium-energy			
Lime	3–5		
Clay bricks and tiles	2–7	2	3
Gypsum plaster	1–4		
Concrete:			
In situ	0.8–1.5		1.2
Blocks	0.8–3.5		
Precast	1.5–8		
Sand–lime bricks	0.8–1.2		
Timber	0.1–5		0.7[b]
Low-energy			
Sand, aggregate	<0.5		0.1
Flyash, volcanic ash	<0.5		
Soil	<0.5		

[a] More complete data is available in the reference cited.
[b] Local air dried.

The embodied energy in the services systems – mechanical, electric, above-ground soil and waste, rainwater disposal and drainage – obviously varies with the building type but services might account for approximately 5–10% of the embodied energy.

Environmental concerns about global warming have meant that the carbon dioxide production associated with energy use has received increasing attention. There are a number of ways in which to reduce the embodied energy and CO_2 production of buildings. The first is to select lower energy materials, paying attention to possible reduced energy substitutes for traditional solutions. A second is to design for longevity. This has a number of aspects, an important one of which is to design buildings of excellence that are acclaimed publicly as they are likely to be well maintained and be in existence for many years. Other aspects include high-quality, durable materials and design solutions that reduce the need for refurbishment. A third aspect is economic use of materials and designing to reduce waste. Recycling of materials is a final and very important aspect. Some materials such as lead have traditionally been recycled;

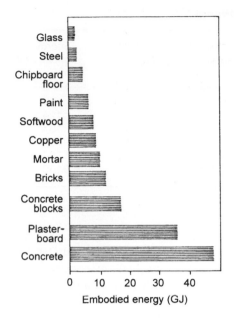

6.1 Approximate embodied energy inputs for a detached house (based on reference 21)

indeed, as long ago as 1775 Samuel Johnson castigated (unjustly, it appears) the clergy of Lichfield for selling the cathedral's lead roof.[22] More recently, efforts have been made to increase the recycling of materials such as concrete and plastics.

6.7 Key materials

The selection of key materials and components will depend on a wide variety of factors: noise transmission, structural spans, fire considerations and cost are just a few of the constraints that guide the building process. In the sections below a brief look is taken at a number of key elements. The information needs to be examined critically and incorporated cautiously into any design. Interactions are numerous: for example, spaces with high exposed ceilings of concrete incorporate a fair amount of energy for a given floor area and the volumetric cost of the space is high; however, the increased mass may eliminate any need for air conditioning and the increased height can allow greater use of daylighting. Victorian era schools of heavyweight construction (little else was available) had high ceilings and large glazed areas to maximise daylight since artificial lighting was rare and expensive. Heating was minimal because it was expensive and pupils and staff were deemed to be robust. The solution had a high embodied energy cost but a low running energy requirement.

Structural materials

Structures obviously need stiffness and a common measure for this is the somewhat surprisingly named elastic modulus, E. Table 6.3 relates stiffness and energy – a higher value of E means a stiffer material.

Table 6.3 Energy requirements and stiffness [24]

Material	Elastic modulus, E (MN/m²)	Density (kg/m³)	Energy (kJ/kg)	Energy cost of one unit of E
Timber (sawn)	11 000	500	1 170	53
Mass concrete	14 000	2 400	720	124
Brick	30 000	1 800	2 800	167
Reinforced concrete	27 000	2 400	8 300	738
Steel	210 000	7 800	43 000	1 598
Aluminium	70 000	2 700	238 000	9 180

The lower energy cost for timber explains in part why there is a growing recognition that wood is an energy-efficient material. The Timber Research and Development Association (TRADA) has examined different methods of constructing a three-bedroomed detached single family house and concluded that a standard timber-framed wall requires 7450 kWh whereas a lightweight concrete block wall requires 12 816 kWh, or 1.7 times as much energy.[23] The figures are quoted cautiously, since it is not always evident that manufacturers – or researchers – make true comparisons. For example, timber elements may not have the same acoustic performance as concrete.

For large buildings, common options are structural masonry, steel frame and concrete frame. Proponents of concrete construction have argued that it requires less embodied energy than steel[25] and the steel industry's view is that overall there is little difference between steel frame and concrete frame.[26] More precise studies of individual buildings are needed to differentiate among the options.

Insulation

Insulation levels have been a prime area of development in energy conservation for several decades. The flow of heat through a wall depends on the insulation level and Appendix B gives a sample calculation. However, the insulation level is only one factor in the energy consumption of a building. Others include how the occupants use the building, how well the controls are understood and how well they work and how much air enters the building. The need for proper sealing is being recognized with legislation requiring pressure testing for larger buildings.

With regard to insulation the basic questions are: What should it be? Where should it be placed? And how much is needed? Masonry construction with a cavity remains important in UK structures. At the De Montfort University Queens Building (Chapter 12), completed in September 1993, 190 mm block (used for structural reasons), 100 mm of Rockwool cavity batts completely filling the cavity and 100 mm external brick, gave a U-value of 0.30 W/m² K.

At the BedZED Development, 300 mm of mineral wool was used between a brick outer skin and a block inner skin to give a U-value of about 0.1 W/m² K. (See Appendix B for a typical calculation method for U-values.)

The principal insulation materials currently available and their characteristics are shown in Table 6.4, and a useful overview of insulation materials can be found in reference 27.

Debate on the choice of insulation materials tends to be related to environmental considerations, durability and buildability. If we first consider the environmental aspects (and keep in mind that it can be misleading to talk about an element out of context, i.e. insulation without considering the entire, say, wall construction) a principal concern for plastic insulants is ensuring that ozone-depleting chemicals are not used in their manufacture. The situation is in flux and so individual manufacturers must be contacted on their products.

For fibrous insulation materials such as mineral wools, concern had been expressed in the past about their possibly being carcinogenic. However, in 2001 a working group of the International Agency for Research on Cancer determined that the more commonly used insulation materials including rock wool and glass wool are 'not classifiable as carcinogenic to humans'.[31]

With regard to embodied energy, it can be seen from Table 6.4 that there is a wide range of values. Insulation materials derived from mineral fibres tend to require less embodied energy than a number of others; similarly, they have lower CO_2 emissions. Sheep will be pleased to learn that their wool, particularly if used locally, has a very low embodied energy. There are, however, large discrepancies in the literature between embodied energy for similar materials and so careful consideration of the validity of data is required. There is general agreement, however, that the use of insulation saves many times its embodied energy – manufacturers' figures range from the hundreds (200 for, say, expanded polystyrene) up to a thousand times (for, say, glass fibre). CO_2 emissions follow a similar pattern of major savings.

Turning to buildability, some architects are concerned about friability of fibrous materials and it is important to ensure that COSHH Regulations are met during installation of such products. Many, however, find that rockwool products suit their designs very well and specify it regularly. Housebuilders are sometimes said to prefer

Table 6.4 Insulation date[a]

Material	Thermal conductivity (W/m K)	Density (kg/m³)	Thermal resistivity (m K/W)	Embodied energy (kWh/m³)
1. Expanded polystyrene slab	0.035	25	28.6	1126[b]
2. Mineral wool quilt	0.040	12	25.0	231[b]
3. Mineral wool slab	0.035	25	28.6	231[b]
4. Phenolic foam board	0.020	30	50.0	1126[b]
5. Polyurethane board	0.025	30	40.0	1126[b]
6. Cellulose fibre	0.035[c]	25	28.6	133[b]
7. Wool	0.037[b]		27.0	31[b]

[a] Data from reference 28 unless otherwise noted.
[b] Data from reference 29. Values in the literature vary widely and will depend on manufacturing techniques, transportation, etc. Caution is advised. Consult with individual manufacturers for more detail.
[c] Data from reference 30.

extruded polystyrene foams because of the ability to lap boards and maintain a clear cavity and good water resistance.

Glazing

Glazing depends on high-energy materials but provides the priceless possibility of views and the more easily priced potential for passive solar gain and daylighting.

Embodied energy figures tend to increase with complexity of design. Wooden-framed double glazing might provide a baseline and as one moves towards argon-filled, sealed double-glazed low-emissivity coating units in aluminium frames with thermal breaks the energy cost rises.

Services

Underground drainage often consists of clay pipes with plastic connectors; above ground soil and waste pipework has moved (quickly) away from traditional cast iron to plastics. Rainwater disposal components tend to be plastic or made of metals such as aluminium. Thus all of these systems have energy implications. However, heating, ventilating, air-conditioning, power and lighting are the main areas of interest.

The best way to reduce the environmental impact of services is to reduce the need for them. Nonetheless, there tends to be an unavoidable minimum input (heating in the depths of winter, lighting at night) and meeting this merits close attention.

Chapter 9 discusses gas and electrical heating ventilation systems and heat recovery.

Pipes are usually in steel or copper and ductwork is mainly in galvanized steel; plastic substitutes for both are becoming more common. All have energy and material implications. Pipe insulants selected should have zero ozone depletion potentials.

Power and lighting cables tend to be in copper with PVC insulation. As discussed previously, the PVC use needs to be kept under review.

Lighting

Environmental concerns focus on the materials in lighting systems and the effects of providing power to the lights. In the UK millions of failed fluorescent tubes are discarded annually. Each tube contains several mg or more of poisonous mercury as well as cadmium, lead, copper, tungsten and a number of other pollutants. The need for safe disposal of lamps is therefore becoming increasingly recognized.

Embodied energy (and CO_2 emissions) associated with lighting systems are low compared with consumption in use.[32] The luminaires (or, less technically, the light fittings – see section 8.3) account for over 80% of the embodied energy in the system and there is scope for lowering the energy content by reduced use of plastics.

Energy in use is, of course, very high. Generating this energy produces CO_2, as we have noted, and a variety of other pollutants: for example, electricity from coal-fired power stations results in emissions of mercury and other trace elements to the air. Thus, even though tungsten lamps do not contain mercury, their use in the UK with its mix of generating stations, including coal-fired, results in emissions of this metal.

A brief discussion of some of the environmental aspects of lighting can be found in reference 33.

Finishes

The primary concerns about finishes have already been mentioned and fall into the two major categories of environment and health.

For floors, denser timber is more hard wearing but, unfortunately, it often comes from tropical rainforests. Again, the recommendation is to ensure that all woods used are from sustainable sources. Where possible, varnishes and paints should be based on water, plant oils or resins and those with fungicides, arsenics, harmful solvents and lead should be avoided.

Floor coverings, if required, should be from renewable materials where possible and, ideally, combine durability with a low embodied energy content. Coverings mainly based on renewable sources include linoleum and cork. Other coverings are often plastics based and may incorporate resins.

For walls and ceilings the first question is whether a finish is necessary. It may be possible to leave brickwork (as at De Montfort's Queens Building (Chapter 12)) or concrete ceilings (as at both De Montfort and RMC (Chapter 11)) free; this can also have thermal advantages in reducing peak summer temperatures.

The disadvantages of commonly-used plasters and renders are that they are from non-renewable resources, require significant energy inputs and are not readily recycled. More environmentally-friendly systems based on alternative renders and silicate paints are becoming available.[34]

Paints for walls and ceilings should meet the same criteria as paints for other surfaces.

6.8 Construction

As lack of space limits discussion of construction issues, we shall only mention a few points.

As insulation levels increase, construction methods change: for example, increased cavity wall insulation is likely to increase cavity width and change such details as the lengths of wall ties.

Also, as insulation is improved, the heat loss through the fabric becomes relatively less important and more attention needs to be paid to ventilation and the concomitant issue of possible condensation.

In our more tightly sealed buildings, condensation can occur in corners, behind doors, on walls and in roofs. Experience has shown that simply providing openings for air is not necessarily sufficient and that air must be able to circulate and take the moisture away. The best approach is 'build tight – ventilate right'. The issues are technical – a brief discussion can be found in reference 35.

Similar problems can occur in refurbishment. Internal insulation of certain concrete system-built flats can result in condensation due to the high water vapour-resistance of the outer leaf. To reduce this risk, cavities can be vented by holes drilled through the outer leaf at high and low level.

To produce energy efficient buildings, cold bridges (local poorly insulated paths from inside to outside) need to be avoided. It is also important to reduce infiltration losses by filling in cracks in walls and ceilings, by sealing holes around pipes, drains and cables and by sealing around window frames with mastic.

On site, inspection is needed to ensure that the insulation and sealing details that have been drawn are achieved.

Our view is that a concept of Planned Cost Effective Adaptation should be developed which anticipates buildings reducing their environmental impact during their lifetime.

Designers and contractors both should ensure that the potential for waste is first reduced and then that any resulting construction and demolition waste is re-used and recycled wherever possible.

The opportunities to reduce waste and achieve more efficient construction systems are factors in the growing interest in off-site fabrication of components varying from bathroom pods to steel frames.

6.9 Evaluation methods

There is a saying that 'comparisons are invidious', and in the environmental field one can add, 'difficult and contentious'. If we first consider materials, current approaches include life-cycle analysis, eco-labelling and ecological footprinting. Life-cycle analyses examine the 'cradle-to-grave' or origin to disposal impacts of materials or products. The analysis of energy used is just one aspect (but perhaps the most easily quantifiable) of such studies which, ideally, should incorporate all environmental effects, in the broadest sense of the term.

Eco-labelling, or environmental labelling, is an attempt to arrive at agreed measures of the environmental suitability of products, which can range from light bulbs to hairsprays to washing machines.

For buildings as a whole there are several schemes and, in the UK, BREEAM (Building Research Establishment Environmental Assessment Method) provides, for a number of building types, including homes,[36] both a useful discussion of the major issues and a ready checklist. Ecological footprinting is a method that attempts to assess the land area needed to produce, say, a newspaper, or more comprehensively the demand that an individual makes on the earth's resources.[37]

Finally, it should be noted that new legislation such as the European Union Energy Performance of Buildings Directive, which requires buildings to be energy-rated, reflects the ever-increasing importance of performance.

Guidelines

1. Use materials with minimal health and safety risks over their life cycle. The use of hazardous materials should be avoided.
2. Avoid HCFCs. Use HFCs only when unavoidable. Use alternatives such as ammonia, hydrocarbons and CO_2 where possible.

3. Investigate the impact of extraction of source materials, pollution associated with manufacturing and disposal and the possibility of recycling materials.
4. Promote sustainability.
5. Embodied energy is important but the greatest savings are to be made from reducing energy in use.
6. Use energy efficient materials. This tends to mean more wood. Restrict the use of plastics and metal to situations where they are indispensable, e.g. copper and aluminium for cables, or where there are significant advantages in weight, strength or durability.
7. Insulate the building fabric well and use high-efficiency glazing systems.
8. CO_2 and other emissions associated with materials need to be considered.
9. The energy and materials implications of services systems have not received sufficient attention. Try to find out more about what will be used in your buildings.
10. The best drawn design can be ruined on site. Attention to the construction process is vital.
11. Evaluation methods are imperfect but are essential for purposes of comparison and useful as checklists.
12. Materials, and our knowledge of them, are constantly changing – keep abreast of recent major developments when specifying.

References

1. Cited in Clifton-Taylor, A. (1972) *The Pattern of English Building*, Faber & Faber, London, p. 6.
2. Littler, J. and Thomas, R. (1979) Thermal storage in the autarkic house. *Transactions of the Martin Centre for Architectural Studies*, University of Cambridge, Cambridge.
3. World Commission of Environment and Development (1987) *Our Common Future*, Oxford University Press, Oxford.
4. See reference 1, p. 294.
5. Baggott, J. (1994). HCFCs declared 'safe' for ozone layer. *New Scientist*, **141**(1911), 15.
6. Anon. (1991) CFCs and buildings. BRE Digest 358. BRE, Garston.
7. Anon. (n.d. – *ca.* 1994) *Saving the Ozone Layer with Greenfreeze*, Greenpeace, London.
8. Anon. (2004) BREEAM Offices 2004. BRE, Garston.
9. Calm, J. M. and Hourahan, G. C. Physical, Safety and Environmental Data for Refrigerants. www.hpac.com/microsites/egb/pdfs/calm–0899.pdf (accessed 14 Feb. 2005).
10. Co-op Bank/Co-operative Financial Services Sustainability Report 2003. www.cfs.co.uk/ sustainability 2003/ecological/conversions.htm (accessed 14 Feb. 2005).
11. Mason, H. J. (1978) *Flint – The Versatile Stone*, Providence Press, Ely, p. 25.
12. Maclaine, D. (1994) Fields of doubt. *Electrical Review*, **227**(10), 28–30.
13. Gimpel, J. (1975) *La Révolution Industrielle du Moyen Âge*, Seuil, Paris.
14. Howard, N. and Sutcliffe, H. (1993) Embodied energy. The significance of fitting-out offices. BRE 180/22/9, BRE, Garston.
15. Ibid., p. 1.
16. Ibid.
17. Spence, R. (1991) *Energy for Building*, United Nations Centre for Human Settlement, Nairobi.
18. Howard, N. (1991) Energy in balance. *Building Services*, **13**(5), 36–8. (NB: Figures in this article should be regarded as approximate as the database was under development at the time.)

19. Halliday, S.P. (1994) Environmental code of practice for buildings and their services. BSRIA, Bracknell.
20. Gartner, E.M. and Smith, M.A. (1976) Energy costs of house construction. Current Paper 47/76. BRE, Garston.
21. See reference 19, p. 36.
22. See reference 1, p. 380.
23. Anon. (1994) TRADA data sheet: timber and brick product energy requirements. TRADA, Rickmansworth.
24. See reference 17, p. 61.
25. Anon. (2000) Ecoconcrete. British Cement Council, Crowthorne, Berkshire.
26. Anon. (1998) Talking steel. *Building Design*, **23**, January, 33.
27. Ramshaw, J. (2003) Insulation choices – embodied energy and performance. *Eco Tech*, **8**, 32–8.
28. Anon. (1995) *The Building Regulations*. L1: Conservation of fuel and power. HMSO, London.
29. Wooley, T. *et al.* (1999) *Green Building Handbook*, E&FN Spon, London.
30. Anon. (1994) *Warmcel Technical Data Sheet*, Excel Industries Ltd, Ebbw Vale, Gwent.
31. Davies, L. (2002) IARC Reclassification of Mineral Wool. Note published by Institute of Occupational Consulting, Edinburgh.
32. Butler, D. and Howard, N. (1992) From the cradle to the grave. *Building Services*, **14**(11), 49–51.
33. Anon. (2004) CIBSE code for lighting. CIBSE, London.
34. Anon. (undated) Natural Building Technologies. NBT Systems Data Sheets.
35. Stephen, R. (2000) Airtightness in UK dwellings. BRE Information Paper 1/00. BRE, Garston.
36. Anon (2004) Eco Homes – The environmental rating for homes. BRE, Garston.
37. Chambers, N. Simmons C. and Wacternogel, M. (2000) *Sharing Nature's Interest: Ecological footprints as an indicator of sustainability*, Earthscan Publications, London.

Further reading

Anderson, J., Shiers, D. and Sinclair, H. (2002) *The Green Guide to Specification*, Blackwell, Oxford.

Anon. (variable dates). BRE Approved Environmental Profile Database. BRE, Garston.

Collins, T. (2000), A call for a chlorine sunset. *Nature*, **406**, 6 July, p.17–18.

Gardner, G. (1998) The insulation cold war. *Building Design*, 17 April, 29–30.

Hall, K. (ed.) (2005) *The Green Building Bible*, Green Building Press, Llandysul.

On-line resources

http://www.constructionresources.com
http://www.ecologicalfootprint.com
http://naturalbuildingproductscouk.ntitemp.com

Energy sources CHAPTER 7

7.1 Introduction

This chapter discusses a range of energy sources, from the traditional to the newer (or rediscovered) alternatives which are beginning to be considered in building applications.

7.2 Energy

Heat and work are forms of energy which are measured in the same units and which can be compared when evaluating different energy conservation options. However, there are important differences: some forms of energy, such as electricity, are readily converted into work or into heat; others, such as fossil fuels, can be converted into heat but only part of the heat energy can provide useful work. The rest is wasted as heat losses in cooling circuits and friction.

The proportion of the energy available as heat that can be converted into work increases as the source of heat gets hotter and the cooling source for rejecting excess heat gets colder. The maximum efficiency of conversion of heat into work is given by:

$$\text{Efficiency} = \frac{\text{Work extracted}}{\text{Heat available}} = 1 - \frac{T_{cold}}{T_{hot}}$$

where T_{cold} is the temperature of the cooling source and T_{hot} is the temperature of the source of heat, both in degrees Kelvin.

(The temperature in degrees Kelvin equals the temperature in degrees centigrade plus 273.)

For example, in electricity production, say, no more than 70% of the fuel burned at 1000 K in an engine with an air-cooled radiator at 300 K can be converted into work ($1-300/1000 = 0.70 = 70\%$). In practice, combustion, friction and generation losses reduce the efficiency and electrical output further. If the temperature of combustion is increased (or the cool temperature is reduced), the efficiency increases – heat is thus more useful, and so more valuable, at higher temperatures.

Primary energy (PE) is that contained in fossil fuels in the form of coal, oil or natural gas or in nuclear energy or hydroelectricity.[1] Delivered energy (DE) is that in the fuel at its point of use after allowing for extraction (or generation) and transmission losses. The portion of the delivered energy which is of benefit after allowing for the efficiency of the consuming appliance is the useful energy (UE). Table 7.1 compares various fuels.

Table 7.1 Energy conversion efficiencies for primary to delivered and delivered to useful energy for space heating fuels

Type of fuel	PE to DE efficiency	DE to UE efficiency	Overall (PE–UE) efficiency
Coal	0.98[a]	0.60[b]	0.58
Gas	0.90[a]	0.70[b]	0.63
Oil	0.93[c]	0.70[b]	0.65
Electricity (grid)	0.32–0.33[d]	0.98[b]	0.29

[a] Reference 2.
[b] Estimates for a variety of devices.
[c] Reference 3.
[d] Reference 4.

Table 7.1 shows that the use of electricity from the grid in low temperature applications such as space heating is efficient at the point of use but very wasteful of primary energy.

In theory, and generally in practice, the 'grade' of energy is reflected in the cost, as energy producers have to pass on to the consumer the costs associated with production, generation and transmission, as well as investment. However, distortions in price result from factors such as government policies and marketing; for example, the costs associated with nuclear power generation and decommissioning are still being debated, while electricity produced by nuclear power is being offered to the consumers at the same rate as that from conventional sources such as gas and coal.

Table 7.2 gives comparative costs for domestic and industrial consumers using various fuels. As can be seen industrial and other large users can use cheaper, lower grade fuels and can negotiate their tariffs and obtain substantially lower unit prices.

Fossil fuels

Fossil fuels are the decomposed remains of plants and organisms resulting from processes dating back to before the dinosaurs. Under the high forces and temperatures that moved and changed the shape of the continents into their present form, these remains fossilized to become today's coal, crude oil and natural gas. Their main components are carbon and volatile hydrocarbons, but they also contain moisture, non-combustible ash and other materials such as sulphur, sodium and nitrogen. Their

Table 7.2 Approximate comparative cost of delivered energy[5]

Application	Coal	GasOil	LPG	Gas	Electricity
Domestic[a]	32[b]	19[b]	124[b]	23	100
Manufacturing industrial[c]	6[b]	65[b]	17[b]	15	38

[a] Average domestic fuel 2003 prices.[6]
[b] Approximate domestic costs based on spot checks. Prices to be confirmed.
[c] Median of prices paid by manufacturing industry, 2003.[7]

properties and compositions vary widely, which has led to classification systems based on factors such as carbon content and calorific value.

Gaseous petroleum fuels

These fuels are low carbon content hydrocarbons (molecules with one to four carbon atoms) which are gases at normal ambient temperatures. They are extracted from underground formations, separated from crude oils during extraction or manufactured from coal. The most common is natural gas which is used generally in the UK for heating and cooking. It consists almost entirely of methane (one carbon atom), has very few impurities and has a high calorific value. Natural gas liquefies at about $-160\,°C$ to form liquefied natural gas (LNG). In this form it can be transported in bulk more easily. Other gases, less commonly used in the UK, are:

— *Town gas* This is manufactured from coal and although once common, has now been superseded by cleaner, higher calorific content natural gas. Town (or manufactured) gas is approximately 30% methane, the rest being a mixture of other gases such as high carbon content hydrocarbons, carbon monoxide, carbon dioxide and hydrogen.
— *Liquid petroleum gases* (LPGs) These are hydrocarbons that are gaseous at ambient temperatures but can be liquefied at moderate pressures. Propane (three carbon atoms) liquefies at 5–15 bar and butane (four carbon atoms) liquefies at 2–6 bar. These gases are used in bottled form as an alternative to natural gas in areas outside the gas mains distribution network. LPGs are heavier than air and leaks tend to settle in lower areas, creating a possible hazard.

Liquid petroleum fuels

These fuels are hydrocarbons of medium carbon content which are liquid at ambient temperatures. The most common liquid petroleum fuels used in building applications are:

— *Paraffin* This is a light distillate heating oil used in small, free standing, flueless, vaporizing heating appliances.
— *Kerosene* This is a distillate heating oil for use in vaporizing or atomizing flued domestic heating boilers, as an alternative to gas.
— *Gas oil* This is a distillate heating oil for larger atomizing domestic and commercial burners.

There is also a whole range known as fuel oils which covers light to heavy residual or blended oils used in large industrial or commercial boilers with large pre-heated storage and handling equipment.

Coal

Coal is classified according to its non-volatile carbon content and calorific value. In general, older coal has a higher carbon content and fewer impurities such as non-combustible ash and sulphur (and therefore has a higher calorific value) than more recent coals. Coal is also graded (and priced) according to size and the extent to which it has been washed in order to remove dust. The more common types of coal in building are:

- *Anthracite* This is premium coal with a high calorific value, lower humidity and low impurities. It is particularly useful in heating applications.
- *Smokeless fuels* These are manufactured fuels produced by heating coal (such as anthracite) in the absence of air.

Other types of coal include lignite, which is of lower calorific value, and peat, which is a much more recent product formed from partly decomposed plant residues. Table 7.3 gives data for common fossil fuels.

When fossil fuels are burned, the products released into the atmosphere include carbon dioxide (CO_2), nitrogen oxides (NOx) and sulphur dioxide (SO_2). The by-products will obviously vary with the fuel.

As we have seen in Chapters 2 and 3, CO_2 emissions are an important concern because of their contribution to global warming. Table 7.4 provides some relevant data.

The emissions from electricity generation reflect the fuel mix and the efficiency of the plant used. In the UK, for example, between 1987 and 2003 there was a marked shift from older coal fired plant to more modern plant using cleaner gas. During that period generation by gas stations increased from about 1% to 32%, and coal reduced from about 70% to 38%. The Electricity Association predicted a reduction in CO_2 NO_x and SO_2 emissions as a result of this shift.[9] Ideally, fuel cost should take into account long-term environmental effects but at present this rarely occurs.

Table 7.3 Approximate properties of typical fuels[8]

Item	Net calorific value (MJ/kg unless indicated otherwise)	Comments on impurities (% kg/kg unless indicated otherwise)
Gas		
Natural gas	35[a]	approx. 0.001% sulphur[b]
Manufactured gas	19[a]	approx. 0.001% sulphur[b]
Liquid petroleum gases		
Propane	46	less than 0.02% sulphur
Butane	46	less than 0.02% sulphur
Liquid petroleum fuels		
Paraffin	44	less than 0.04% sulphur
Kerosene	44	less than 0.2% sulphur
Gas oil	43	less than 0.2% sulphur
Coal[c]		
Anthracite	29–32	less than 1.1% sulphur

[a] MJ/m^3.
[b] By volume.
[c] Coal used for power generation is usually lower grade. The 1990–2003 average gross calorific value of coal used in UK power stations was 25–26 MJ/kg.

Table 7.4 Relationship between primary fuel use and carbon dioxide emission in the UK [10]

Fuel	CO_2 emission (kg/kWh delivered)
Electricity (grid)	0.52
Coal (anthracite)	0.32
Coal (domestic)	0.29
Fuel oil	0.28
Gas oil	0.27
Liquid petroleum gas (LPG)	0.22
Other petroleum gas (OPG)	0.21
Natural gas	0.19

7.3 Combined heat and power

In main thermal power stations some 32–46% of the fuel burned is converted into electricity,[11] the remainder is dissipated as waste heat in chimneys, cooling circuits and in the generator itself. Power stations are normally sited far from densely populated areas and so the waste heat is lost. Further losses are incurred during transmission to the points of use.

In combined heat and power (CHP) or co-generation installations, gas or diesel internal combustion engines (or, in large installations, gas turbines with reheat) drive a generator and produce electricity. Typically 23–28% of the fuel used can be converted into electricity in this way. Of the other 72–77%, some 63–66% is available as waste heat suitable for normal space and hot water service heating so that most of the energy in the fuel is useful, as shown in Figure 7.1.

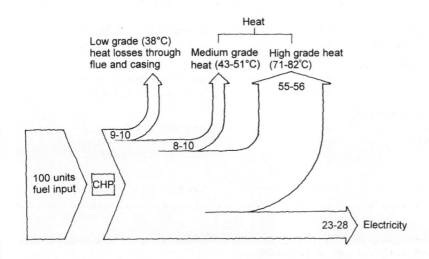

7.1 Typical energy balance of a CHP unit.[12]

These systems were developed from traditional standby generating sets, but there is now considerable expertise in gas-fired CHP units with an electrical output of 15–300 kW running at the same time (i.e. in parallel) with the normal electrical supply. In this arrangement electricity from the grid provides any demand in excess of the capacity of the CHP and absorbs any spare capacity. The heat from the cooling circuits is used for space and hot water service heating and is treated similarly to a conventional boiler. Figure 7.2 shows a typical unit with the panel removed and on the right the actual installation with an acoustic enclosure at the De Montfort University Queens Building (Chapter 12). Other options such as multiple (and/or larger) units, emergency operation, and other fuels are also possible.

Some points to be considered when planning an installation are:

- High overall efficiencies can only be achieved if both heat and electricity are used. The unit(s) must be run for long periods to be cost-effective (say 3500–4500 hours per year), thus, the unit must be relatively small or excess heat will need to be rejected. Consequently, CHPs are most suitable for providing a base load.
- It is possible to sell surplus electricity to the utilities or to pay for using the grid to transport surplus from one site to another. However, price per unit of electricity exported to the grid is approximately half of the price for imports (see Appendix D for further details). This favours smaller CHPs with all output used on site to displace imports at a higher tariff.
- Maintenance and service costs are significant and must be fully considered, allowing for regular replacement of components and major servicing over the lifetime of the installation.

7.4 Heat pumps

(a) Typical unit.
(b) Installation at De Montfort.
7.2 CHP installations.

A heat pump is a device that transfers heat from a cold source to a hot one. Heat travels from the hot to the cold and work must be expended to reverse the flow of

(a)

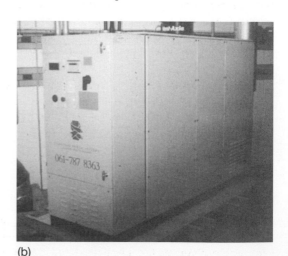

(b)

heat. This work is dissipated as useful heat and adds to the heat extracted from the cold area.

A domestic refrigerator is similar in that heat is removed from the inside of the refrigerator, thus cooling it, and is rejected to the room which is warmer. In a heat pump the emphasis, however, is normally on obtaining the maximum amount of heat for the warm room.

The performance of a heat pump is usually indicated as the coefficient of performance (COP), which is the ratio of the useful heat delivered to the warmer area (Q_{hot}) to the work expended in the process (W). When the temperature differential between the cold and the hot sources increases the heat pump has to work harder to transport the heat and the COP diminishes. The maximum COP that can be achieved is given by:

$$\text{COP} = \frac{Q_{hot}}{W} = \frac{T_{hot}}{T_{hot} - T_{cold}} \qquad \text{where } T_{hot} \text{ and } T_{cold} \text{ are in degrees K.}$$

For example, a heat pump may extract heat from ambient air at −3 °C to deliver warm air at 47 °C. The refrigerant will be colder than the cold source and hotter than the hot source. For example, a heat pump may extract heat from ambient air at -3 °C to deliver warm air at 47 °C. The refrigerant will be, say 15 °C colder than the cold source and 15 °C hotter than the hot source so that the heat pump operates internally between –18 °C and +62 °C. In the example the maximum COP is about 4.2. In practice, mechanical losses etc. reduce the seasonal COP typically to 2.5–3.5; that is to say, for each unit of energy used to drive the heat pump, 2.5–3.5 units of useful heat are delivered.

Heat pumps in buildings are usually powered electrically, and some of the energy savings will be lost during electrical generation and transmission. They can also be driven by oil- or gas-fired engines, but this is more applicable to larger installations.

While this may seem attractive, there are further considerations, which mean that the environmental 'acceptability' of heat pumps is not evident. Firstly, the higher CO_2 production of electrically-run heat pumps in comparison with, say, a gas-fired condensing boiler (Table 7.4) would, of course, be a disadvantage; the higher monetary cost of electricity may also mean that heat pumps are not economic. Another factor is the type of refrigerant used. Chlorofluorocarbons (CFCs) are now prohibited and the less environmentally damaging hydrochlorofluorocarbons (HCFCs) are being phased out. Chlorine-free HFCs and alternative refrigerants such as ammonia (NH_3), hydrocarbons (HCs) and CO_2 are being developed although there are safety issues to be resolved with some.

7.5 Renewable sources of energy

Over 70% of fuel used for electricity generation and nearly 90% of the energy supply in the UK is derived from solar energy in the form of fossil fuels, formed by photosynthesis, accumulated, concentrated and stored over millions of years. The size of the reserves may be debatable, but this energy use represents a constant depletion of resources and at some future time this source will be exhausted.

Renewable sources, all ultimately derived from the Sun's energy, include solar power (both active, i.e. incorporating moving water or air, and passive systems and photovoltaic devices), wind and wave power and biological sources such as wood and fuels derived from crops. They are renewable (or, similarly, sustainable) because the Sun will continue to provide their energy.

However, the energy density of solar and wind energy is not great and the availability is very variable. To overcome these disadvantages, a combination of research, product development, legislation and financial incentives is required. In England and Wales licensed electricity suppliers have been required to supply a proportion of their electricity from renewables. The target was set at 3% in 2000 and annual targets updated to reach 10% by 2010.[13]

Active solar heating

Active solar heating uses collectors to convert solar radiation into heat for space and hot water heating. Collectors are surfaces painted black to absorb most of the incoming radiation, glazed and insulated to reduce the heat losses and suitably orientated (usually within 30° or so of due south) to optimize the amount of energy incident on them.

The heat generated in the collectors is removed and circulated by a water circuit (or less commonly air). A typical installation usually also includes some form of heat storage such a hot water store, as illustrated in Figure 7.3. There is also likely to be a means of providing back-up heating when solar heating is insufficient. There are many types of collectors for different applications. Where heat at low temperatures

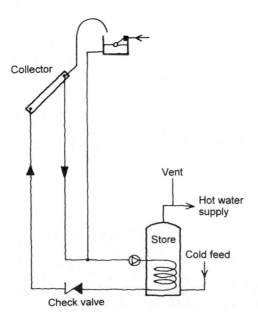

7.3 Simplified schematic of a typical solar water heater system (indirect, vented, pumped circulation).[14]

is adequate, such as for swimming pools, the collectors can be simple, unglazed, un-insulated, and therefore relatively cheap. Where the heat is needed at higher temperatures, such as for space heating, or where collectors are required to operate at lower ambient temperatures, the heat losses become greater and the collectors must be insulated and glazed. In more advanced collectors the heat losses are further reduced by special surface treatments known as selective coatings that reduce reradiation from the collector while retaining a very good absorption of solar radiation. Even lower heat losses and higher temperatures can be achieved by enclosing the collectors in a vacuum (evacuated tubular collectors) and, beyond that, by focusing incident solar radiation into a smaller area, as in concentrating collectors.

Some 20–50% of the monthly solar radiation incident on a collector can be delivered as useful heat output from solar water heating applications in the UK. This amounts to some 720–1800 MJ/ m^2 per year;[16] the precise figure depends on factors such as type of system, as well as water and ambient air temperatures. There is an increasing interest in solar heating applications but their use in middle and northern European climates needs to be considered carefully. Swimming pool heating and solar hot water heating are, however, becoming established in sunnier climates and where traditional fuels are unavailable or expensive.

Figure 7.4 shows a system at an alligator farm in Florida, where solar-heated water (at a temperature of 60 °C) is blended with well water at 22 °C which is required for washing down the buildings that house the alligators.[17] By maintaining a constant 32 °C for the buildings and all water used, the alligators experience no thermal shock and grow very well.

7.4 Solar water heating at a Florida alligator farm.[15]

Photovoltaics

When solar radiation (commonly known as sunlight or more simply but somewhat misleadingly as light) falls on certain materials known as semiconductors, the energy of the incident radiation (photons) frees some of the electrons in the materials. A difference in potential develops within the material thus providing a direct conversion of energy into electricity. These photoelectric cells are typically manufactured from silicon, grown either in single crystals which improves the performance at increased costs or, more commonly, as a polycrystalline or amorphous silicon deposit onto a substrate, which gives lower efficiencies but much lower costs. The photoelectric cells are then interconnected in groups with diode protection from the environment. The power produced by a PV system is directly related to local solar radiation levels which are of course subject to the vagaries of the weather as well as daily and seasonal cycles. Therefore standalone systems usually incorporate batteries, a charge controller and an inverter, if AC output is required. Photovoltaic energy is well established in remote areas without access to the grid for high value applications such as beacons, communication and specialist equipment. There is growing interest in grid-connected systems which do not normally have batteries. With the resulting increase in production and improvements in technology their cost effectiveness is likely to continue to improve. Output varies, but figures of 250–350 MJ/ m^2 per year have been measured in the UK. Chapters 14, 17 and 18 discuss PV installations and Appendix D gives further details.

Wind power

Air motion requires energy: the Sun and the rotation of the Earth provide the energy to move large air masses, thus producing wind, and this 'wind' energy can be harnessed by sails to move a boat or to turn a windmill to pump water or grind flour, or, more recently, generate electricity.

The maximum power that can be derived from a wind turbine is given by an expression derived from Betz.[18] Aerodynamic, mechanical and electrical losses will reduce this further.

$$P = \eta \times 0.35 \times (A \times u^3)$$

where P is power output (W), η is the plant efficiency which is in the range of 50 to 70% to account for mechanical and electrical losses, A is the area swept by the wind turbine blades (m^2), and u is the wind speed (m/s).

Wind speed varies constantly in magnitude and direction. This is normally presented as a frequency distribution curve, which indicates the percentage of the time that a given wind speed is exceeded. Since economics dictate that equipment must operate during most of the year to repay the investment, most windmills are designed to operate at common wind speeds of 5–15 m/s (Appendix A). Batteries (or the grid) provide power when speed is low and safety devices protect the mill in high winds.

Typical wind speeds in the UK are about 5 m/s (see Appendix A). Higher wind speeds, although less frequent, carry more energy and therefore the effective speed is higher. The ratio between the effective speed and the wind speed exceeded 50% of the time varies with the site. Here, in both instances, a factor of 1.4 has been used. Winds are also slowed by friction with the ground and deflected by the topography and surrounding buildings. Table 7.5 illustrates these factors and compares the estimated output of a wind turbine at a site of an urban inland location with an exposed site near the coast.

Wind generation from medium to large wind turbines is now an established technology. Between 1999 and 2003 onshore and offshore installed capacity in the UK more than doubled and the main barrier to future developments is likely to be in planning consent.

The principal applications are:

– remote locations without access to the grid (i.e. where alternative sources of electricity are equally costly and/or restricted), particularly for small essential loads (e.g. communications equipment), or where the vagaries of the wind are not critical (e.g. base load lighting), and
– large installations in exposed locations where economies of scale can be applied (i.e. wind farms, as shown in Figure 7.5).

The current estimated generation costs from wind turbines are similar to domestic tariff rates but a higher price is often set as an incentive to investors. Chapter 15 discusses the windmill installation at the Millennium Centre, Dagenham and Chapter 17 discusses wind power at Beaufort Court, King's Langley.

Hydroelectric power

Hydroelectric power relies on the energy released when water moves from high to low levels, much as wind power relies on air movement from regions of high to low pressure. However, the energy density of hydroelectric power is greater than that of the wind because of the higher density of water (1000 times greater than air), the higher pressures available, and the concentrating effect of water in rivers and lakes. Consequently, hydroelectric power generation has been a conventional source of

Table 7.5 Approximate comparison of power from wind turbines and the effect of exposure[19,a]

	Wind speed[b] (m/s)	Effective speed[c] (m/s)	Terrain correction factor	Adjusted effective speed (m/s)	Estimated electrical output at the adjusted effective speed[d] (W/m^2)
Urban location	4.0	5.6	0.6	3.4	10
Exposed location on coast	5.5	7.7	1.0	7.7	110

[a] Assumed height of 10 m.
[b] Typical wind speed exceeded 50% of the time.
[c] This is the wind speed which, if constant, would provide the same amount of energy over the course of a year as the actual varying wind speed.
[d] Estimated from the Betz formula assuming an annual plant efficiency of 65%.

7.5 Wind farm.[20]

energy for some time; water power for milling has, of course, been used for much longer.

In the UK, hydroelectric generation accounts for under 2% of the total electricity supply; most comes from large-scale sites in Scotland. The potential of the unexploited resources is limited by economic and environmental constraints, but it is estimated that the generating capacity of sites in excess of 5 kW could be increased by 50%.[21]

With developments in low-cost controls and the relative increase in the cost of energy, mini- and micro-hydroelectric generation is becoming increasingly viable, and it is now easier to exploit previously uneconomic water courses with smaller flows or falls. In the short to medium term, however, micro-hydroelectric generation in the UK will remain an option only for private schemes for small communities isolated from the grid and near to water courses. The technology is well established, however, and the potential for other parts of the world is promising.

Biomass

Wood was once the traditional fuel for heating and cooking, and is still widely used in many parts of the world. In many countries it was replaced when more efficient, controllable, convenient and less locally polluting forms of heating became available. Modern wood-burning stoves, however, can be efficient and control has improved. Small installations have been traditionally manual, but automatic fuelling is becoming available for smaller units burning woodchips, pellet fuels or similar. Larger installations can use a wider range of fuels and have automatic fuel loading and removal of ashes and mechanized fuel deliveries.

Agricultural waste, such as straw bales from cereal crops, and forestry waste, such as branches and tree tops from thinning and timber harvesting, can also be used as fuels. Because the calorific value is low they are an economic alternative only where they are locally available at low cost; sometimes they are used in dual-fuel installations with, say, oil. Typical data is given in Table 7.6. Wood pellets, made of compressed wood particles have a higher calorific value and can be transported economically over larger distances.

As we have seen, in photosynthesis green plants use sunlight to convert carbon dioxide in the atmosphere (or dissolved in water) into oxygen and fixed carbon. When the carbon compounds are burned, the chemical energy in the compounds is released and the carbon is emitted as carbon dioxide, so that over the lifetime of the plant there is no net increase in carbon dioxide. For energy crops, selected and planted to provide fuel, the annual energy yield available is estimated at 130–700 GJ/ha.[23] Worldwide, an estimated 5.5 million hectares are expected to be surplus to requirements for food production by the year 2010,[24] and some of this area could be used for energy crops.

Other

Fuel cells are a technology which generates electricity directly from a chemical reaction. Depending on the fuel used, emissions at the point of use are very low. With hydrogen in particular the by-products are only water and heat. Hydrogen and methanol powered fuel cells are attracting a lot of interest from car manufacturers as an alternative to hydrocarbon fuels. Hydrogen can be produced by electrolysis using energy from renewable sources provided the generation capacity can be expanded to meet the demand. There are some pilot projects for fuel cell CHP applications in buildings but there is still requirement for further development. Many are optimistic that hydrogen will form the basis for an environmentally friendly future.

Electricity generation from offshore underwater tidal currents and from waves is also being developed and tested. Other sources, including geothermal energy, biogas and municipal waste, may have a greater role to play in the future.

Guidelines

1. The use of high-grade fuels such as electricity in low-grade applications – for example, heating – is wasteful of primary energy.

Table 7.6 Typical characteristics of dry agricultural waste fuel[22]

Item	Net calorific value (MJ/kg)	Comments on impurities
Harvested straw	15[a]	15% moisture content
Timber harvest waste	10[b]	55% moisture content; ash 1–2% kg/kg; low sulphur content
Miscanthus[c]	17	2.7% ash

[a] At 15% moisture content.
[b] At 55% moisture content.
[c] Typical bioenergy crop, see Chapter 17 for further details.

2. Fuels have impurities that become pollutants, so fuels must be chosen carefully.
3. Combined heat and power (CHP) can be very energy efficient but must be judged on its merits for each application.
4. There is an increasing interest in solar heating applications but their use in middle and northern European climates needs to be considered carefully.
5. There is growing interest in photovoltaics as the technology develops and prices fall.
6. Wind power is now technically proven and likely to increase in use but public concerns over the environmental impact of onshore applications need to be addressed.
7. Biomass application schemes are promising where not too far from the fuel source.

References

1. Bakke, P. *et al.* (1975) Energy conservation: a study of energy consumption in buildings and possible means of saving energy in housing. BRE. Current paper 56/75. BRE, Garston.
2. Shorrock, L.D. and Henderson, G. (1990) Energy use in buildings and carbon dioxide emissions. BRE, Garston.
3. See reference 1.
4. Anon (2004) Fuel input generation for electricity generation 1970–2003. DTI, London. www.dti.gov.uk/energy_stats/total_energy/index.shtml
5. Anon (2004) Prices of fuels purchased by manufacturing industry in Great Britain. DTI, London. www.dti.gov.uk/energy/inform/energy_prices/tables_mar04.shtml
6. Ibid.
7. Ibid.
8. Anon. (2001) *CIBSE Guide C: Reference Data*, CIBSE, London.
9. Anon. (n.d., *c.* 1993) *The Benefits of Electricity – An Overview*, Electricity Association.
10. Anon. (2004) BREEAM Offices 2004 Assessment Criteria. BRE, Watford
11. See reference 9, p. 14.
12. Forrest, R., Heap, C. and Doggart, J. (1985) Small-scale combined heat and power. *Energy Technology*, Series 4. Energy Efficiency Office, Harwell.
13. Anon. (2003) Yearly Renewables Targets. DTI, London. www.dti.gov.uk/energy/renewables/policy/yearly_targets.shtml
14. Wozniak, S. (1979) *Solar Heating Systems for the UK: Design, Installation and Economic Aspects*, Department for the Environment Building Research Establishment. HMSO, London.
15. Healey, H.M. (1994) A modular solar system provides hot water for alligator farm. *ASHRAE Journal*, **36**(3), 53–5.
16. Anon. (1989) Solar heating systems for domestic hot water. BS 5918:1989. British Standards Institution, London.
17. See reference 13.
18. Golding, E.W. (1955) *The Generation of Electricity by Wind Power*, E & FN Spon, London.
19. Anon. (1988) *CIBSE Guide A2: Weather and Solar Data*, CIBSE, London.
20. Anon. (1994) *British Wind Energy Association*, BWEA, London.
21. Anon. (1994), *An Assessment of Renewable Energy for the UK*, Energy Technology Supply Unit. HMSO, London.
22. See reference 21.
23. See reference 21.
24. See reference 21.

Further reading

Allen, P. and Todd, B. (1997) *Off The Grid – Managing Independent Renewable Electricity Systems*, Centre for Alternative Technology, Machynlleth, Powys.

ETSU (1985) *Energy Technology Series* No. 5: Heat pumps for heating in buildings. Energy Efficiency Office.

Evans, B. (1998) Revolution waits in the wings. *Architects' Journal*, **207**(5), 40–1.

CHAPTER 8　　# Lighting

8.1 Introduction

As noted in Chapter 3, artificial lighting in an office building accounts for approximately 16% of its energy consumption. For other building types the figures will vary but it is almost always significant.

Substituting increased natural daylighting (daylight includes sunlight, which is the direct beam, and skylight, i.e. visible diffuse sky radiation) for artificial lighting in offices and other buildings offers large potential energy savings. This is true provided overheating and glare can be controlled and provided significant increased heat losses do not result.

8.2 Daylighting

Most people prefer daylight. The contact with changing natural light is physiologically, psychologically and architecturally important. Le Corbusier said 'architecture is the masterly, correct and magnificent play of masses brought together in light . . .'[1]

Daylight availability varies enormously (in this way it is very similar to natural ventilation) and is a key design issue. (Temperature variations, on the other hand, are more seasonal and are therefore easier to control; noise level variability depends very much on the site). The average levels of daylight are given in Appendix A.

The lighting level in the space is very important. One aspect of this is the daylight factor, which is defined as the illuminance received at a point, indoors, from a sky of known or assumed luminance distribution, expressed as a percentage of the horizontal illuminance outdoors from an unobstructed hemisphere of the same sky; direct sunlight is excluded from both values of illuminance.[2]

Recommended daylight factors, as shown in Table 8.1, exist but need to be treated with caution, because they are not in fact high enough if the optimum use of daylight

Table 8.1 Recommended daylight factors[3]

Space	Minimum (%)	Average (%)
Lounges in dwellings	0.5	1.5
School classrooms	2	5
Offices: general	2	5
Hospital wards	1	5

is to be made. The best guidance is probably to say that daylight should be maximized subject to the constraints of glare, increased solar gains and possible greater heat loss. (We shall return to this issue below. Reference 4 states that 'if electric lighting is not to be used during daytime, the average daylight factor should not be less than five percent'. Chapter 18 discusses an example of how such high daylight levels can be provided.) There are also recommended lighting levels for spaces as shown in Table 8.2.

It should be noted that the values given in Table 8.2 are guidelines and judgement should be used. For offices, in particular, there is the possibility of lowering the level to, say, 300 lux, provided task lighting can supply 500 lux (or more) where required.

The amount of light that enters a space obviously depends on the areas and disposition of the glazing. To a large extent the amount of daylight at a point in a room depends on the area of sky that can be seen through the window. Thus, there tend to be wide disparities in natural light levels between areas close to windows and those some distance from them. In Appendix B a very simple calculation for the average daylight factor is given. Numerous more sophisticated calculation procedures[5] exist and computer simulations can now produce very accurate images of internal light conditions.

It is common to apply the daylight factor to what is known as the standard overcast sky illuminance of 5000 lux.[5] This value is exceeded about 85% of the standard working year (Table A.3). Thus, under a standard overcast sky a daylight factor of 10% at a point near a glazed wall would give an illuminance of 500 lux. This indicates that if you had a desk at that position in a general office the light level would normally be sufficient, but for 15% of the working hours you would need to turn some lights on to maintain the recommended lighting level. It also shows why the 'recommended' daylight factors of Table 8.1 are too low. Could the lighting at times be too bright? In

Table 8.2 Recommended lighting levels[6]

Space	Standard maintained illuminance (lux)
Atria	
– general movement	50–200
– plant growth	500–3000
Chemical industry	
– colour inspection	1000
Classrooms	300
Electrical industry	
– assembly work, medium	500
Healthcare, general lighting	500
Libraries, reading areas	500
Offices	
– filing, copying, etc.	300
– writing, typing, etc.	500

one way the answer is 'no' since most people will happily read outside on the sunniest summer day (the danger here being excessive exposure to UV radiation), but there is the possibility of glare, and the very real problem of potential thermal discomfort due to direct solar gain at such a position demands some form of solar control.

Although the main reason to make effective use of daylight is to reduce artificial energy consumption there are also potentially useful heat gains available. A rule of thumb is that an illumination level of about 1000 lux outside would correspond to total solar radiation of about 10 W/m^2 on a horizontal surface. (This is based approximately on the highest daylight level of approximately 100 000 lux, corresponding to about 900 W/m^2 on a horizontal surface.)

Making effective use of daylight depends very much on planning the building. As discussed in Chapter 2, highly articulated spaces with a greater perimeter length will normally offer more potential daylight (and natural ventilation) but at the possible cost of greater heat loss. The art lies in finding the right balance – one key element of which is reducing the heat loss at night (see, for example, Figure 8.3).

The more glazing at the perimeter wall, the higher the daylight factor. Figure 8.1 gives a more quantitative idea of measured daylight factors in three cases.

A key point for all these examples, and for many of the more innovative daylighting systems, is that ceiling reflectances must be kept high by using very light colours (Table 8.3); another consideration is that the ceiling should not be encumbered by bulky artificial lighting systems. High ceilings with windows running up to them, as in Victorian schools and hospitals, help to increase light levels (and produce good uniformity - see below). For very rough calculations, Figure 8.2 gives some rule of thumb guidelines.

Obviously, wherever possible, activities that need a great deal of light should be placed near the perimeter. Traditionally, this has been done in weavers' cottages and

Table 8.3 Approximate values of light reflectance[7]

Material	Reflectance
1. *Internal*	
White paint[a]	0.85
White paper	0.8
Light grey paint	0.68
Strong yellow paint	0.64
Wood – light veneer	0.4
Strong green paint	0.22
Quarry tiles	0.1
Carpet – deep colours	0.1
2. *External*	
Snow (new)	0.8
Portland stone	0.6
Sand	0.3
Brickwork (red)	0.2
Green vegetation	0.1

[a] BS 4800 colour codes are given in the original reference.

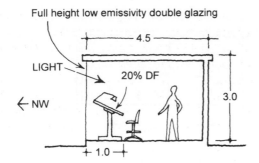

Full height low emissivity double glazing

LIGHT

20% DF

← NW

4.5

3.0

1.0

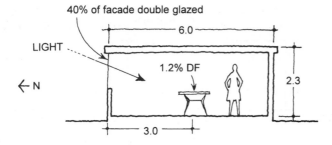

40% of facade double glazed

LIGHT

1.2% DF

← N

6.0

2.3

3.0

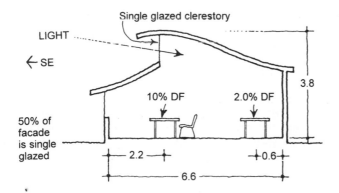

Single glazed clerestory

LIGHT

← SE

10% DF   2.0% DF

50% of
facade
is single
glazed

2.2   0.6

6.6

3.8

8.1 Indicative daylight factors.

large buildings. (In the author's offices – a nineteenth-century former piano factory – glazing occupies 35% of the perimeter wall area and pianos were polished close to the windows in an area with a daylight factor of about 6%.) Similarly, a variety of spaces that need lighting only occasionally (such as storerooms), or that have lower lighting requirements (such as circulation spaces), should be moved towards the interior.

One standard reference suggests that when the average daylight factor exceeds 5% on the horizontal plane an interior will look cheerfully bright, and when the factor is below 2% the interior will not be perceived as having adequate daylight and electric

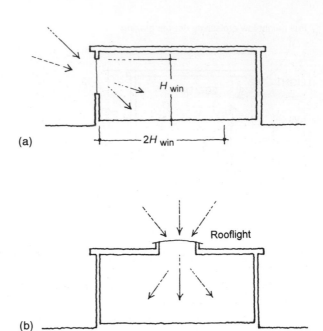

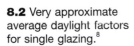

(a) Sidelighting average DF
 $= 20(A_g/A_f)\%$
(in the area adjacent to the
window to about $2H_{win}$
away).
(b) Horizontal skylight:
average DF
 $= 50(A_g/A_f)\%$,
where DF is the daylight
factor as a percentage, A_g
is the area of glazing (m^2),
A_f is the area of floor to be
lighted (m^2), and H_{win} is the
window head height (m).

8.2 Very approximate
average daylight factors
for single glazing.[8]

lighting may be in constant use.[9] Achieving 5% (and, often, ideally more) is by no means easy in many situations and requires careful design.

The quality of lighting and the perceived need to switch on artificial lighting depends much on the range of light levels from the front to back of the space. Too great a range can give a predominantly gloomy character to the space. One criterion[10] for acceptable uniformity in spaces with windows on one wall is that $(d/w + d/h)$ shall not exceed $2/(1 - R_b)$ where d is the depth of the room, w is the width of the room, h is the height of the window head above floor level and R_b is the area weighted average reflectance of the half of the interior remote from the window. For single–storey buildings, or the uppermost floor of multi-storey buildings, rooflights can provide additional lighting away from the perimeter. Figure 8.3 shows an example at a Hampshire school, in which the rooflights were fitted with movable insulated shutters to reduce heat loss at night.

Rooflights increase uniformity and also have the advantage of 'seeing' more of the sky than vertical glazing. The sky is also brighter directly above us than it is at the horizon, and this is a further advantage for rooflights.

On a standard overcast day the approximate lighting levels just outside a vertical window and outside a horizontal skylight are about 2000 lux and 5000 lux, respectively. Thus, it makes sense to consider the use of skylights where possible, always keeping in mind that solar gain needs to be kept under control. Thought also needs to be given to the effects of toplighting because some people find that this gives objects a duller appearance owing to a lack of modelling (see section 8.3 below for a further discussion). Another consideration is that the heat loss of a rooflight on a clear night can be greater than that of a window because the rooflight 'sees' more of the cold sky than the window and so radiation loss is greater; losses due to conduction/convection can also be somewhat greater with rooflights.

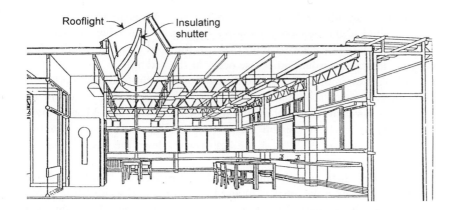

8.3 Rooflights at Crookham Church School. (Architects: Edward Cullinan Architects.)

In more common applications where only sidelighting is possible, one way of attempting to improve uniformity has been the use of lightshelves (Figure 8.4) and other reflective mechanisms such as prismatic glass; a number of innovative daylighting systems are reviewed in reference 11.

The BRE has carried out extensive tests on lightshelves in south-facing walls and found that in sunny conditions they had the advantage of shading an area of the room close to the windows from direct sunlight but resulted in some light loss and provided relatively small redirection of light.[12] In overcast skies light levels are reduced by 5–30%, depending on the position in the room with the 5% loss being at the back of the room. This pattern of some light loss and some redistribution in sunny conditions and light loss in overcast conditions was also found to be valid in the same study for similar techniques such as mirrored louvres and prismatic films.

One point to keep in mind when evaluating such systems is the relative proportion of times when it is overcast and when it is sunny. In the UK it is often overcast – the number of hours of bright sunshine in London averaged over the year is only about four per day and in the winter when light levels are lower the figure is less than half this (Appendix A).

There may, however, be other ways of looking at lightshelves that make them more promising. One possibility is to use them as a perimeter zone for running services.

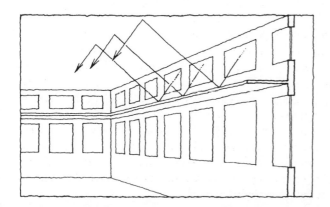

8.4 Typical lightshelf.

Another is to have an adjustable lightshelf similar to a Venetian blind. In sunny conditions the blind would be closed to reflect light up to the ceiling; under overcast skies it would be open allowing more light to the zone near the window (and eliminating the losses that would otherwise occur at the ceiling).

If we now look at attempting to provide daylight or sunlight from rooftop level to floors lower down we find it is not easy. In atria with large glazed areas and few obstructions it is not a problem but shafts are much less effective. Measurements of the internal light levels at the top and bottom of the stacks at De Montfort (Chapter 12) show losses of 98–99% as the light travels down the stack. Figure 8.5 shows this schematically.

The De Montfort shafts are diamond-shaped (3.2 m between the most distant vertices and 1.4 m between the other two) and lined with a dull white fabric (which covers the acoustic attenuation). They were not specifically designed as light shafts and by altering the geometry and using smoother and whiter surfaces reflective losses could be reduced and light levels at the base could be improved somewhat but not greatly.

Larger shafts have the disadvantage of requiring more space for openings on intermediate floors. A common problem with all shafts is that with time the surfaces become less clean and less reflective. Additionally, the light that eventually reaches the space may have a curiously dead quality even on sunny days because the direct sunlight is effectively transformed into diffuse light by reflections.

More highly technological systems of light pipes or piped sunlighting with slow-tracking lenses or mirrors exist and others are being developed. However, it must be noted that in many areas of the world, including much of northern Europe, systems designed particularly for direct sunlight are not likely to be appropriate as sunlight is often in short supply. Furthermore, the systems cannot enhance the diffuse light and, indeed, reduce it. It would seem, therefore, that these solutions with their higher costs are destined for a minority of buildings in sunnier climates.

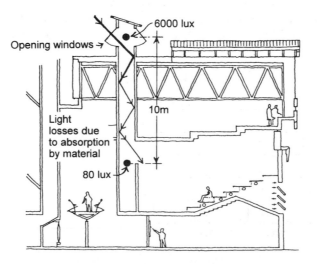

8.5 Representative measured light levels in tall shafts at De Montfort University's Queens Building.

8.3 Artificial lighting

Daylighting and artificial lighting

The artificial lighting scheme and daylighting should complement each other. One important aspect of this is the quality or spectral composition of the light, which varies with a number of factors including the position of the Sun and weather conditions, the most obvious being cloud type and cover.

In order to allow discussion and comparison of light sources, the concept of a correlated colour temperature (CCT) has been developed. The CCT can be considered, somewhat simplistically, to relate the temperature of a radiating body or object and the colour of the light it produces. (Fuller discussions can be found in reference 13.) By international agreement the CCT of average daylight (sunlight plus skylight) outdoors is taken as 6500 K (in practice it varies between 4000 and 12 000 K).[14]

The CCT allows light sources to be located on a simple warm–cold scale. If the CCT is less than 3300 K the source is considered warm; if it is between 3300 and 5300 K it is intermediate; and if it is above 5300 K it is cold. If this seems paradoxical, consider the warm glow from a dying open fire compared to the 'white' heat from much hotter temperature sources such as molten metal.

It is, of course, very difficult to obtain quantitative levels of natural daylighting with artificial sources (although in hospital operating theatres local light levels can reach 100 000 lux) but for our normal activities this is fortunately not necessary. Another question, however, is: How close can we come to reproducing the quality of daylight? In Figure 2.4 we saw a typical distribution spectrum for solar radiation at the Earth's surface but, of course, solar radiation and daylight vary continuously. The CIE (Commission Internationale de l'Eclairage, or, in English, the International Commission on Illumination) has defined a set of reference illuminants mainly based on a series of spectral power distributions of phases of daylight.[15] The CIE colour rendering index (CRI) is a measure of how accurately colours of surfaces illuminated by a given light source match those of the same surfaces under one of the reference illuminants. The closer the CRI is to 100, the better the agreement.

Figure 8.6 gives spectral power distribution chart (i.e. the amount of light generated in each band) and the CRI for a tubular fluorescent lamp. The shape of this

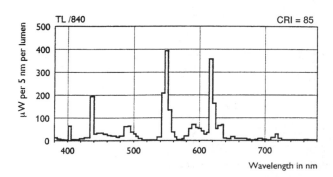

8.6 Spectral power distribution chart.[16]

diagram can be compared with that of north sky daylight, as shown in Figure 8.7 (for interest, the distribution is also shown in Figure 2.4).

What can be done to ensure that daylight and artificial light complement each other, or are at least mixed successfully, always remembering, of course, that the spectral composition of the daylight reaching indoors will be a function of the glazing, the colours of the interior, the shape of the room and so forth? As a simple experiment, switch on a tungsten table lamp (CCT of 2700 K) on a white tablecloth in broad daylight and note the yellow glow that contrasts with the cool daylight. A number of architects favour lighting that is warm but not too yellowish. The CIBSE way of expressing this is their recommendation that, in general, room lighting discrepancies between the colour of electric light and daylight can be reduced by using lamps of intermediate colour temperature (3300–5300 K).[17] Creating transition zones from mainly naturally lighted areas to more artificially lighted ones is part of the art of architecture.

A second issue in artificial light complementing daylight is how light can model objects, i.e. show their texture and form. A completely uniform distribution of light will not reveal these features. Again a simple experiment is to view an object in a conservatory and then in a room indoors. Perhaps the lighting that we find most natural corresponds to daylight through a window and so comes in at an angle of, say, 10–60° from the horizontal. Narrow beam intense light sources can create sharp-edged dark shadows (observe how, in some museums, overhead lighting causes the top part of the picture frame to shade part of the painting). For a 'softer' lighting scheme, the light needs to be diffused in some way.

A more technical discussion of this would refer to the vector/scalar ratio. Put simply, the scalar illumination is that arriving at a point irrespective of direction and the vector illumination is that from the strongest source (a detailed explanation can be found in reference 18). The vector/scalar ratio represents the strength of the flow of light. Ratios of 1.2–1.8 are satisfactory in normal general lighting where perception of faces is important;[20] a ratio of 1.0 is a soft lighting effect and one of 3.0 gives strong contrasts. Daylight can produce similar effects and if it comes in at low angles can cause objects in the space to be silhouetted. The art is again to combine daylight and artificial light to produce an enjoyable environment.

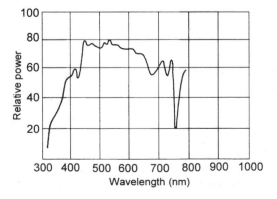

8.7 Spectral composition of north sky daylight of 5700 K (*NB*: infrared not shown).[19]

The practical importance of the above considerations is that unless daylight and artificial light work together the artificial light will not be turned off and so there will be no scope for energy savings.

Another growing concern among designers is lighting and health. SAD (seasonal affective disorder) is thought to be caused by a lack of sunlight and experiments are being undertaken to see if the use of more 'natural' indoor light sources will increase the sense of well-being and health.

Light sources

The prime measure of efficiency for a light source is the ratio of the light emitted to the power input, expressed as lumens/watt. Manufacturers are constantly attempting to increase this ratio and designers are trying to keep abreast with corresponding low-energy schemes that use less power to provide the same illuminance, i.e. lumens per square metre of area lighted, expressed as lux. For office design an achievable rule-of-thumb objective is that the installation should not exceed 2.5 W/m^2/100 lux. (More detailed information is given in reference 21).

Table 8.4 gives data on eight principal light sources used throughout the fields of domestic, commercial and industrial lighting.

An important development is the growing availability of LEDs (light emitting diodes) for both internal and external applications. These can offer potential energy savings, long lifetimes (in excess of 50 000 hours) and good colour rendering (CRI of 90, and over) and will play a major role in the future.

Luminaires

The luminaire is the apparatus that controls the distribution of light from a lamp or lamps and that incorporates fixings and the components to connect the lamp to the power supply. One principal reason for controlling the light emitted is to avoid glare.

As has been said, glare is like intelligence – easy to recognize, but difficult to define. Essentially, glare is an imbalance between the general brightness and any particular source of light. Glare can result naturally, as when direct sunshine strikes a desk, or from artificial light sources. With glare from sunshine the danger to energy use is that blinds, or some other shading devices, will have to be used and artificial lighting increased. This 'blinds down, lights on' syndrome is by no means uncommon. It is often exacerbated by the blinds being left down even the following day which may, in fact, be cloudy.

Two kinds of glare are commonly referred to. Disability glare impairs the ability to see detail and discomfort glare causes visual discomfort.

A detailed technical discussion of designing to avoid glare can be found in reference 22, but here we shall simply consider some broad design approaches. The standard technique for reducing glare from lamps is to incorporate some kind of diffuser or screen in the luminaire. Another technique is to recess the lamp or treat it in some other way so that most of the light falls in a fairly narrow beam. This can result in a rather gloomy effect if extreme care is not taken.

A fairly common solution that has appealed to many architects has been uplighting. Particularly in tall spaces where the ceiling has architectural interest, uplighting has

Table 8.4 Light source data[a]

Light source	Approximate (lm/W)	Approximate colour temperature (K)	Indicative colour rendering index	Approximate lamp life (h)	Capable of being dimmed	Starting	Comments
1. General lighting service GLS (filament)	10–20	2700	100	1000	Yes	Prompt	Energy inefficient; produces a great deal of heat
2. Tungsten halogen (TH) (filament)	10–20	2700–3050	100	1500–5000	Yes	Prompt	
3. Tubular fluorescent (MCF)	60–100[f]	3 categories a. Less than 3300 b. 3300–5300 c. Greater than 5300	50–98[d]	10 000–20 000	Yes	Prompt	Controls incorporating high frequency ballasts improve the energy efficiency of fluorescent lamps
4. Compact fluorescent[b]	50–80[f]	2700–6500	82–98[d]	10 000–15 000	Yes	Variable	
5. Metal halide[b]	80–90	3000–10 000	60–93[d]	5600–13 000	Limited dimming available	1–2 minute run-up	
6. High pressure mercury (MB[b])	40–60	2700–4000	42–52[d]	14 000–28 000[c]	No	2–5 minute run-up	
7. High pressure sodium (SON[b])	60–120	2000–3000	25–80[d]	14 000–28 000[f]	Limited dimming available	1.5–6 minute run-up	
8. Low pressure sodium (SOx)	100–190	2000–2300	[e]	11 500–23 000	No	8–12 minute run-up	

[a] Compiled principally from reference 23.
More complete and detailed information may be found in manufacturer's literature.
[b] Numerous designations exist. See reference 24.
[c] Blended high pressure mercury lamps have shorter lamp lives.
[d] See reference 25 and manufacturers' literature.
[e] Only yellow light is produced and no colour rendering is provided.
Mainly used for road lighting.
[f] Variable – consult manufacturers' literature.

been able to highlight features and provide a glareless, generally soft appearance at lower levels. The difficulty from the point of view of energy conservation is that uplighting is inherently inefficient because the reflectance from the ceiling is, at best, likely to be 85% with a bright, freshly painted white surface. Recently, designers and manufacturers have been developing combined up- and downlighters to try to improve the overall lighting efficiency and still achieve the desired aesthetic effect.

8.4 Controls

An underlying concept in thinking about controls and, indeed, buildings is zoning. Very broadly, a space will have a perimeter zone and a core zone. As the building becomes more complex it may have corridors, entirely enclosed internal spaces and atria.

Zones obviously have varying requirements. At the perimeter, control of solar radiation is most important and we have seen some solutions in Chapter 4. It is also at the perimeter that there is normally the greatest possibility of using daylight efficiently. This can be done by arranging for lights serving the perimeter to be controlled separately from others in the space. A more sophisticated (and more costly) system is to use photocells to control the lighting in a zone that can receive both daylighting and artificial lighting. As the daylight level increases the artificial lighting level is lowered (and vice versa) very gradually.

Zoning is also important in allowing for different use patterns in a space. For example, in an open plan office zoning could provide a background security level of illumination throughout the space for those working late, and local controls could increase the light levels where people were actually working.

Controls can provide a much needed (and appreciated) element of personal influence over one's working environment. In many ways this is similar to the thermal issues discussed previously. The ability to control the lighting level and direction contributes to the perception of an enjoyable environment.

One approach is to be able to dim the lights as this can help with certain glare problems as well as reduce energy consumption. Dimmers for tubular fluorescent lights have been available for some time and manufacturers are striving to develop systems for other high-efficiency light sources.

A second very flexible approach is to provide background lighting to, say, 200–300 lux and then use adjustable personal task lights to increase this level as necessary. Energy efficiency is, of course, improved if these lights are turned off when they are not needed.

A technical issue for controls is how quickly light sources will respond. With several high-efficiency lamps the run-up times tend to mean that lights are left on when not required rather than switched off and switched on again when needed.

Generally, controls must be understandable. For example, switches need to be readily accessible and it should be clear what lights they serve. For many buildings, attractive schematics providing information on the controls adjacent to the switches would be very useful. Controls have been a nexus of constant conflict in the past because of conflicting requirements among individuals and between individuals and automatic control systems. 'Fiddling' with the controls is a fine art in many buildings.

More sophisticated controls incorporated in building management systems may offer reasons for optimism. Already, systems that shut off lights automatically at the end of the working day are providing energy savings, and devices such as movement detectors that ensure that lights are on in particular spaces, such as WCs, only when needed are proving useful.

One can envisage intelligent controls that 'correct' or at least warn users of the results of their decisions. For example, if a user overrides a control system that has switched off the perimeter lights on a bright sunny day, a computer could flash a message every ten minutes giving the increased energy use, CO_2 production and likely temperature rise resulting from the action. It might even, after a period of time, override the user, requiring the perimeter lights to be switched on again if desired.

The scope for control is enormous – in some buildings every single fitting is addressable and can be controlled – but, as always, the balance between simplicity and complexity will determine the overall success of any design.

Finally, any energy-efficient scheme must consider access to the controls and the lights themselves. Maintenance is a particular issue. Fittings that look slightly dirty may be wasting about 30% of their light and surveys often show greater losses. But, obviously, to clean them they must be readily accessible.

Guidelines

1. Most people prefer daylight – make effective use of it.
2. Ensure that the average daylight factor is adequate.
3. Ensure that there is sufficient uniformity of daylight.
4. Daylighting systems do not increase the amount of light – they simply redistribute it.
5. Direct sunlight from all azimuths and altitudes must be considered.
6. Consider use and maintenance.
7. Use high-efficiency fittings but ensure that they are appropriate to the application.
8. Use appropriate control systems.

References

1. Le Corbusier (1927) *Towards a New Architecture*, transl. by F. Etchells, The Architectural Press, London.
2. Anon. (2004) *CIBSE Code for Lighting 2004*, CIBSE, London.
3. Anon. (1987) *Window Design: CIBSE Applications Manual*, CIBSE, London.
4. Anon. (1992) Lighting for Buildings: Part 2: Code of practice for daylighting. BS 8206. British Standards Institution, London.
5. Anon. (1986) Estimating daylight in buildings. BRE Digest 309. BRE, Garston.
6. Anon. (2002) *CIBSE Code for Lighting*, CIBSE, London and accompanying CD-ROM.
7. See reference 4, p. 27.
8. Brown, G.Z., Haglund, B., Loveland, J., Reynolds, J.S., and Ubbelohde, M.S. (1992) *Inside Out: Design Procedures for Passive Environmental Technologies*, Wiley, New York.
9. See reference 3, p. 10.
10. See reference 4, p. 19.

11. Littlefair, P. (2000) Developments in innovative daylighting. BRE Information Paper 9/00. BRE, Garston.
12. Aizelwood, M. (1993) Innovative daylighting systems. An experimental evaluation. *Lighting Research Technology*, **25**(4), 141–52.
13. See reference 6.
14. See reference 2.
15. Anon. (1988) *Method of Measuring and Specifying Colour Rendering of Light Sources*, CIE Publication 13.2. Commission Internationale de l'Eclairage, Paris.
16. Anon. (2003) *Philips Lighting Catalogue 2003/4*, Philips, Croydon.
17. See reference 6.
18. See reference 6.
19. See reference 4.
20. See reference 2.
21. See reference 2.
22. See reference 2.
23. See reference 2.
24. See reference 2.
25. See reference 2.

Further reading

Aizlewood, M.E. (1998) Interior lighting: a guide to computer programs. BRE Information Paper 16/98. BRE, Garston.

Littlefair, P. and Aizlewood, M.E. (1998) *Daylight in atrium buildings*. BRE Information Paper 3/98. BRE, Garston.

Slater, A.I., Heasman, T.A. and Bordass, W.T. (1996) People and lighting controls. BRE Information Paper 6/96. BRE, Garston.

CHAPTER 9

Engineering thermal comfort

9.1 Introduction

Engineering thermal comfort is a broad topic. In this chapter we shall examine the range of environmental conditions encountered and look at the heating, ventilating and air-conditioning systems that are available to designers. Great emphasis is placed on natural ventilation systems because of their growing importance.

9.2 A range of conditions

Heating to cooling covers a broad field and the role of ventilation varies with the position in that field. Figure 9.1 shows the principal considerations.

9.3 Heating

Energy sources have been discussed in Chapter 7. For space heating, gas, if available, is the most common choice. Coal is not often specified at present as gas is a cleaner option. Our discussion will therefore concentrate on gas and, to a lesser extent, electricity.

[a] This is obviously variable since, during a hot day, one may want to restrict the amount of fresh air to avoid bringing in outside air that may be warmer than the air inside.
[b] Fresh air alone, or, more commonly, a mixture of fresh and recirculated air, can be cooled below the ambient temperature. The latter is normally more energy efficient.

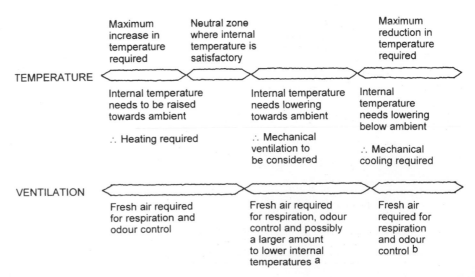

9.1 Heating, cooling and ventilation considerations.

TEMPERATURE

Maximum increase in temperature required

Neutral zone where internal temperature is satisfactory

Maximum reduction in temperature required

Internal temperature needs to be raised towards ambient

∴ Heating required

Internal temperature needs lowering towards ambient

∴ Mechanical ventilation to be considered

Internal temperature needs lowering below ambient

∴ Mechanical cooling required

VENTILATION

Fresh air required for respiration and odour control

Fresh air required for respiration, odour control and possibly a larger amount to lower internal temperatures [a]

Fresh air required for respiration and odour control [b]

The standard 'wet' or hydronic heating system consists of one or more boilers (or combustion devices), pumps, heat emitters, interconnecting pipework and a control system, as shown in Figure 9.2.

Gas-fired boilers can be grouped into the three categories of conventional, high-efficiency and condensing. Conventional boilers keep initial cost down by such factors as simplicity of design and reduced boiler insulation levels. High-efficiency boilers have more efficient heat exchangers and better casing insulation. Condensing boilers have an additional or enlarged heat exchanger which, under all conditions, recovers sensible heat from the flue gases (Figure 9.3) and, under suitable conditions, recovers latent heat from the condensation of water vapour (generated during combustion) in the flue gases. The formula for the combustion process for natural gas (which is principally methane) is:

$$CH_4 + 2O_2 \longrightarrow CO_2 + 2H_2O$$
Methane   Oxygen   Carbon dioxide   Water

For latent heat to be recovered, the return water temperature must be below the dewpoint of the flue gases. Removal of heat from the gases results in lowered temperatures, which are typically in the range of 40–80 °C.[2]

Figure 9.4 shows the effect of return water temperature on a condensing boiler with flue gases with a dewpoint of about 54 °C. (See reference 3 for a detailed explanation of this.) As the return water temperature falls below the dewpoint, latent heat is recovered and the boiler efficiency increases. When condensing boilers first started to become commercially available in the early 1980s, designers considered oversizing the radiators to reduce the return temperature, thus encouraging condensation of flue gases at the boiler. This was soon found to be uneconomical, however, and radiators are now sized in the normal way.

The growing general recognition of the environmental benefits of condensing boilers has led to legislation encouraging and in some cases requiring their use. The

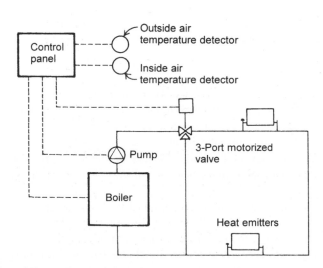

9.2 Conceptual schematic of a heating system.

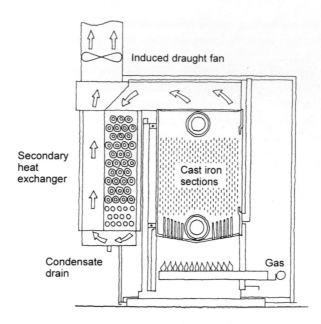

9.3 Broag gas-condensing boiler.[1]

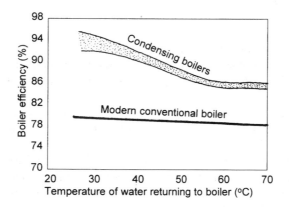

9.4 Effect of return water temperature on boiler efficiency.[5]

present market for single boiler installations is for a high–efficiency boiler which will operate as a condensing boiler in some conditions (this will depend in part on the boiler controls). In multi–boiler installations it is common to combine a high–efficiency condensing boiler with high–efficiency boilers.

The SEDBUK (Seasonal Efficiency of Domestic Boilers in the UK) provides data on a range of small–scale boilers.[4] Seasonal efficiencies of over 90% are given on 'A' rating.

Figure 9.5 gives very indicative efficiencies as a function of load.

CO_2 emissions and other pollutants can also be reduced in other ways and boiler and burner manufacturers are striving to develop appropriate products. The basic reaction for burning natural gas has been given above. Stoichiometric combustion is when the fuel is reacted with exactly the amount of oxygen to oxidize, in our example,

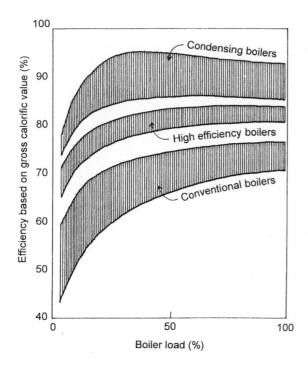

9.5 Typical efficiency ranges for boiler types at part load.[6]

all the carbon and hydrogen in the methane to CO_2 and H_2O. Thus in stoichiometric combustion there is no incompletely reacted fuel and no unreacted oxygen. In practice this is not attainable, and normally excess oxygen or excess air is required in the combustion process.

The more efficient the burner, the lower the amount of excess air used and the lower the volume of CO_2 emissions in the flue gases. Burners are available commercially that operate at less than 5% excess air.

Manufacturers of combustion equipment are attempting particularly to reduce pollution by lowering NOx emissions (Chapter 2). NOx are formed during combustion by reactions of nitrogen and oxygen at high temperatures and by oxidation of organic nitrogen in (certain) fuel molecules.[7] Part of the effect is related to excess air because NOx levels peak when excess air is at 40% and reduce significantly as the excess air decreases.[8] The main part, however, concerns control of the burner flame temperature. Methods of reducing NOx emissions include burner adjustment, flame inserts and staged combustion and delayed mixing.[9]

A bewildering range of units exists for expressing NOx emissions and care must be taken in comparing different manufacturers' products. A convenient unit is mg NOx per kWh of delivered energy. Figure 9.6 indicates some typical NOx emissions from different types of boilers.

We have seen in Chapter 2 that SOx (SO_2 and SO_3) are air pollutants. SO_2 is produced when sulphur or sulphur-containing fuels (such as coal) are burned. As SO_2 is the result of the fuel, sophisticated burner techniques have little effect on reducing its formation. SO_3 is created in the atmosphere by the oxidation of SO_2 under the influence of sunlight; some can also be produced directly in the combustion process.

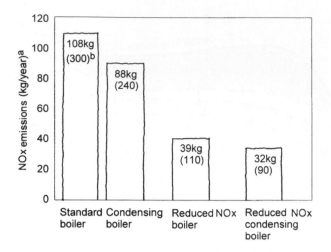

^aBased on a 300 kW boiler operating for 1200 hours per year.
^bFigures in parentheses are mg NOx per kWh of delivered energy.

9.6 NOx emissions from different boiler types, with and without reduced NOx burners.[10]

It is interesting to note that clients with important building portfolios are now including in their environmental policy statements specific directives to designers to minimize SO_2 and NOx emissions to the atmosphere when designing and installing new equipment. (For existing plant the same clients require a high standard of maintenance to help reduce these emissions.)

9.4 Heating: distribution systems and heat emitters

Hydronic heating systems (i.e. those containing water) with gas-fired boilers are the most commonly installed. Pipework tends to be steel in larger installations and copper in smaller and domestic buildings. Energy losses associated with distribution can be reduced by locating the energy sources near the centre of the load. Careful attention to insulation of pipework – ensuring that energy is delivered only at the emitters – is another way to lower energy consumption. A range of typical emitters and their approximate net outputs is shown in Figure 9.7.

The choice of which emitter to use is often related to issues of appearance and space. In terms of energy efficiency an important consideration is the pattern of occupation of the building. The thermal response of the building fabric is a vital consideration also. For intermittently occupied spaces a rapid thermal response is required from the heat emitters so that the occupants do not have to wait long to be comfortable. In continuously occupied buildings this is less important. Another consideration is that if casual heat inputs from, say, solar gains or a crowd of people increase quickly, one wants the heat emitters to respond by decreasing their outputs to avoid overheating. There is thus an interaction between heat emitters and their controls. Table 9.1 gives an approximate grouping of the thermal response of some heat emitters.

Warm air heating systems have been in common use in American homes for many years. Because suburban bungalows often have basements, ductwork can easily be run

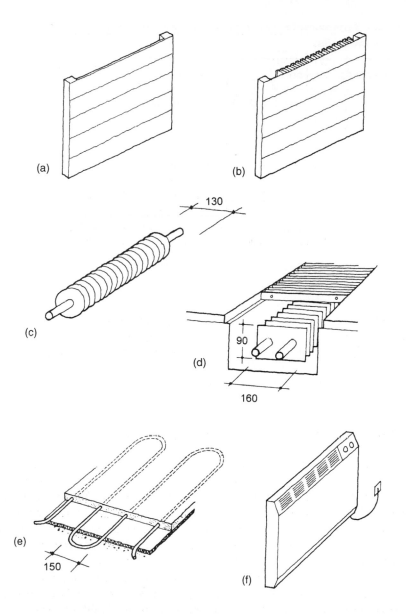

(a) Panel radiators *ca.* 1300 W/m^2
(b) Panel radiators with fins *ca.* 2500 W/m^2
(c) Spiral-gilled tubes *ca.* 600 W/m^2
(d) Perimeter trench heater *ca.* 1200 W/m^2
(e) Underfloor heating *ca.* 90 W/m^2
(f) Electric convector heater *ca.* 5600 W/m^2

9.7 Typical heat emitters and approximate heat outputs.

between basement ceiling joists and taken up to wall registers that are manually controlled. In the UK, with space at a premium, fewer warm air systems have been installed, either in domestic or commercial buildings.

Interest has grown of late because of the possibility of incorporating heat recovery on the extract air. This has occurred simultaneously with improved building insulation and so the ventilation heat loss has become more important in relation to the fabric loss. Figure 9.8 shows a typical domestic warm air heating installation.[11] Ventilation and heat recovery are discussed in more detail below.

Electric heating from the present mix of fuels feeding the national grid is not the best environmental choice (Chapter 7). Nonetheless, in particular circumstances

Table 9.1 Approximate thermal response of heat emission methods

Thermal response category	Heat emitters
Rapid	Forced warm air, unit heaters, fan convectors
Intermediate	Natural convectors, radiators, perimeter heaters, spirally wound tubes
Slow	Underfloor heating

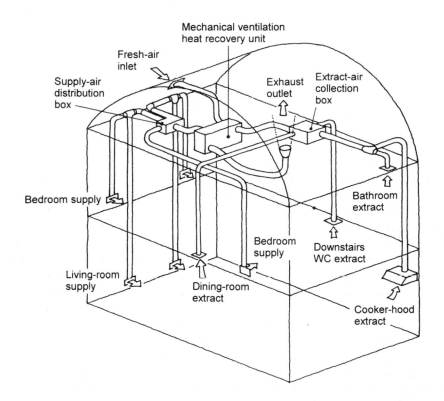

9.8 Domestic warm air heating system.

electric heaters are used and a range of emitters is available. Instantaneous heaters include the old-fashioned radiant bar elements, electric panel heaters (Figure 9.9) and compact heaters with fan assistance to increase heat output.

Over the past two decades, many developers (particularly in the residential sector) have favoured electric heating systems over gas fired systems. Installation and maintenance costs are generally lower and electric heating systems also avoid the complexities associated with meeting the legislative requirements for gas installations (for example, ensuring a ventilated route for the supply pipe).

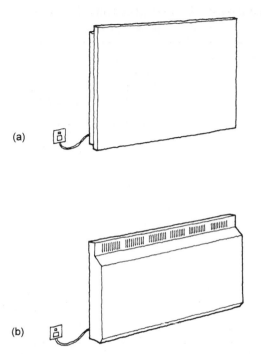

(a) Panel heater
(b) Storage heater

9.9 Electric heaters.

However, public policy as represented, for example, by the proposed changes to Building Regulations, aims to greatly reduce the heating demand of new buildings; this is part of a wider strategy to cut carbon emissions in the UK in accordance with the Kyoto protocol. The method for demonstrating compliance with the Regulations involves making an assessment of the carbon emissions of any proposed building. Given that the carbon dioxide emissions associated with electricity supplied from the grid are more than twice that associated with gas for each kilowatt-hour of delivered heat, potentially expensive compensatory measures will need to be included in electrically heated buildings if they are to comply with the proposed legislation.

The environmental argument against electric heating is reinforced by the Government's Housing Energy Efficiency Best Practice Programme which recommends that electric heating should be considered only in limited circumstances.[12]

Heat pumps are one form of electrically powered heating that can be considered as relatively environmentally friendly when compared to resistive types of electrical heater due to their high coefficient of performance (COP) (see Chapter 7). This is recognized by the Building Regulations. However, most available heat pumps generally make use of refrigerants which have a global warming potential[13] and once installed there is the possibility that they will be used for cooling in the summer resulting in further energy consumption.

An increasing proportion of electricity is being generated from renewable sources and consumers are able to purchase exclusively 'green' electricity from specialist suppliers. However, this is not currently recognized as a means for overcoming the legislative restrictions placed on electric heating. Nonetheless, there may come a point in the future when electricity from renewable energy sources is recognized as a suitable, practical environmental solution for all energy demands. Part of this debate will be about whether the 'high-grade' energy of electricity is better used for power than heating.

9.5 Ventilation

The requirements for ventilation have been described in Chapter 2. In large buildings, ventilation played a number of roles. The nineteenth- and twentieth-century urban atmosphere was often extremely polluted by the products of combustion, especially from coal, and engineers and architects of the time began to investigate ways of producing cleaner air inside their buildings.

One of the landmarks of twentieth-century modernism, Frank Lloyd Wright's Larkin Building (Figure 9.10), was designed as a sealed box with mechanical ventilation to combat pollution (and probably noise) from the adjacent railway.

The development of highly mechanically serviced buildings in the twentieth century can be explained by a number of factors, foremost among which was cheap energy – the running costs for buildings were not sufficiently high to encourage investigation of alternative ways of servicing buildings. (To an extent this is still true and the major impetus at present for more progressive policies is environmental, i.e. global warming and depletion of the ozone layer rather than cost.) The pristine glass block was also a potent symbol of architectural modernity and social progress. These buildings were seen as prestigious, and an alliance of clients, architects, structural engineers, mechanical engineers and suppliers of everything from ventilation fans to cladding systems collaborated to produce them. In many ways these systems were effective. Filters allowed them to remove many outdoor pollutants, ductwork systems with attenuators were able to reduce external noise significantly and cooling

9.10 Larkin Building by Frank Lloyd Wright.[14]

permitted comfortable temperatures to be maintained in the hottest conditions. The price has proved to be too high, however, and we need to determine the most appropriate way to provide comfortable temperatures while simultaneously addressing related issues such as air quality and noise.

A possible starting point is vernacular architecture. What kinds of solutions were developed from generations of practical experience and immediate observations of the effects of design variations? Figures 9.11 and 9.12 show details from English maltings; Figure 9.13 shows a Scottish whisky maltings; and Figure 9.14 a nineteenth-century American home with a cupola (also known as a lantern, belvedere or observatory).

These buildings first of all strike us by their character and point us towards an architecture that can combine functionalism, fantasy and a sense of place. If we look at them more technically we note that the areas for air to enter and leave are large. In the maltings air comes in through the large louvred windows and passes up through the flooring blocks, then out through the kiln-vent. (Temperatures are, of course, higher than most of us would find comfortable.) In the American house the cupola was centred over the staircase and by opening windows and internal doors on all levels ventilation occurred from side to side and bottom to top. Protection against rain is provided by louvres in the kiln-vent in the maltings and windows in the house. Air paths are simple and direct.

If we now consider our naturally ventilated homes we know that they are usually ventilated from one side (Path 1) or cross-ventilated (i.e. ventilated from one side to the other) (Path 2). Figure 9.15 shows this and additionally has a third path which is stack effect ventilation.

The natural driving forces for ventilation are the wind and stack effect. These are, of course, variable and less easy to control than the forces associated with

9.11 Maltings kiln vents at Langley, Worcestershire.[15]

9.12 Maltings air inlet detail at Mistley, Essex.[16]

9.13 Scottish whisky maltings at Dalwhinnie.[17]

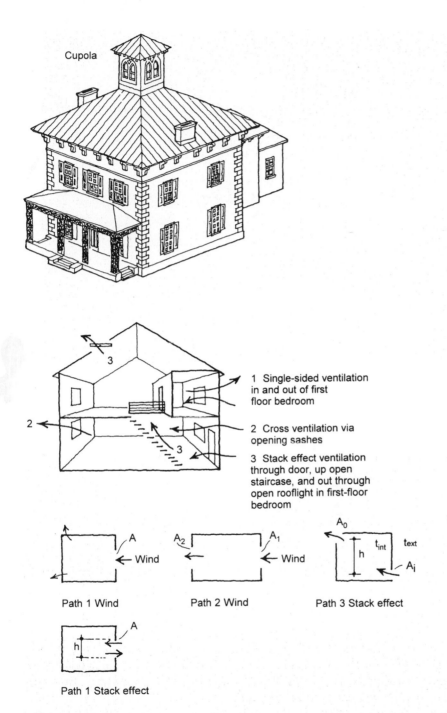

9.14 Nineteenth-century American house.[18]

1 Single-sided ventilation in and out of first floor bedroom

2 Cross ventilation via opening sashes

3 Stack effect ventilation through door, up open staircase, and out through open rooflight in first-floor bedroom

Path 1 Wind

Path 2 Wind

Path 3 Stack effect

Path 1 Stack effect

9.15 Ventilation paths in a simplified and modified Victorian terraced house.

mechanical systems. On the other hand, where used successfully they can enormously reduce capital and running costs of mechanical and electrical plant and reduce the need for plant space. Figure 9.16 shows indicative flow patterns over a simple building.

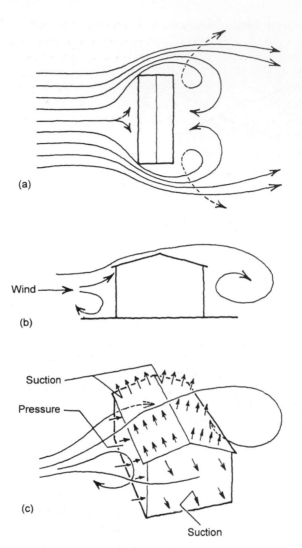

(a)

Wind →

(b)

Suction —

Pressure —

(c)

Suction

9.16 Schematics of air flows and resulting pressures on an isolated building.[22]

The driving force of the wind, and hence its ability to ventilate a building, is related to its velocity pressure, VP, which is given by:

$$VP = 0.63 \; u^2$$

where VP is in pascals (Pa) and u is the speed of the wind (m/s).[19] Thus if the wind speed is 2 m/s, the velocity pressure is about 2.5 Pa and if the speed is doubled, the velocity pressure becomes 10.5 Pa. These pressures are very low compared to those encountered in mechanical ventilation systems.

Very approximately the pressure on the windward face of a building is about 0.5–0.8 times the velocity pressure of the wind, and on the leeward side the negative pressure is about 0.3–0.4 times the same pressure. The total wind pressure acting across a building is thus approximately equal to the wind velocity pressure. (More detailed discussions of these issues are found in reference 20.)

The wind speed at Heathrow Airport in London exceeds 1.6 m/s about 80% of the time; at lower heights the wind speed is reduced and in more urban areas it is further reduced. The conclusion is that in many situations wind cannot be relied upon to provide any significant driving force. (The coincidence of a critical situation of low wind speeds and high temperatures is discussed in Appendix A.) Another point to note is that because the wind speed is so variable, the ventilation rate of naturally ventilated buildings changes more than in a tightly controlled mechanically ventilated construction.

Returning to Figure 9.15, let us examine each path quantitatively.

Path 1 The amount of ventilation due to the wind alone is given by:[20]

$$Q = 0.025 \, Au$$

where Q is the volume flow rate (m³/s), A is the area of the opening (m²) and u is the wind speed (m/s).

Assuming $u = 2$ m/s and $A = 1$ m², we have

$$Q = 0.05 \, \text{m}^3/\text{s}$$

Path 1 can also have a stack effect. The term refers to the process of heated air (which is, of course, less dense) rising in a space. Air inside most of our occupied buildings tends to be warmer than outside air. It rises, finds its way out through an available opening and is replaced by cooler external air at low level.

The stack effect equation is rather more complicated:[21]

$$Q = 0.2A \left[\frac{(t_{int} + 273) - (t_{ext} + 273)}{t_{mean} + 273} \, gh \right]^{0.5}$$

where t_{int} is the internal temperature (°C), t_{ext} is the external temperature (°C), t_{mean} is the mean of t_{int} and t_{ext} (°C), g is the gravitational constant (9.81 m/s²) and h is the height of the opening (m). (This formula assumes a discharge coefficient for flow through the window of 0.61.)

Assume $A = 1$ m², $t_{int} = 20$ °C, $t_{ext} = 15$ °C and $h = 1$ m and we have

$$Q = 0.08 \, \text{m}^3/\text{s}.$$

Path 2 Cross ventilation by the wind is given by

$$Q = 0.61 \, A_w u \, (\Delta C_p)^{1/2}$$

where A_w is the equivalent area (m²) for wind-driven ventilation, given by:

$$\frac{1}{A_w^2} = \frac{1}{A_1^2} + \frac{1}{A_2^2}$$

and ΔC_p) is the change in the pressure coefficient. ΔC_p ranges from 0.1 for a sheltered site to 1.0 for an exposed one. Again we have assumed a discharge coefficient through the window of 0.61.

If we take A_1 as 1 m^2 and A_2 as 1 m^2, u as 2 m/s and $\Delta C_p = 0.5$, we have

$$Q = 0.62 \text{ m}^3/\text{s}.$$

This shows the commonly appreciated advantage of cross ventilation over single sided ventilation.

There is also stack effect ventilation associated with Path 2 but we shall not cover it here.

If both wind and stack effects operate at the same time, the behaviour is complex and depends on the relative values of the opening areas and the relative values of the wind and stack effects. A reasonable approximation of the combined flow rate is simply to take the larger of the two separate values. Note that the resultant ventilation is not the sum of the two.

Path 3 Here we see a more pronounced stack effect ventilation path. The basic pressure difference equation for stack effect ventilation is given in a simple form by:[23]

$$\Delta p = 3462h \left(\frac{1}{t_{\text{ext}} + 273} - \frac{1}{t_{\text{int}} + 273} \right) \text{Pa}$$

where h is the vertical distance between the inlet and outlet (m), t_{ext} is the external temperature ($^\circ$C) and t_{int} is the internal temperature ($^\circ$C). For common temperatures encountered in building design an approximate form of this equation is:

$$\Delta p = 0.043h \, (t_{\text{int}} - t_{\text{ext}}) \text{ Pa}.$$

Thus, if h is 6 m and the temperature difference is 5 $^\circ$C, the pressure difference is 1.3 Pa. Again, this pressure difference is very low compared to those encountered in mechanical systems.

The volume flow rate is given by:

$$Q = 0.827 \, A \, (\Delta p)^{0.5}$$

where

$$A = \frac{A_i \; A_o}{\left(A_i^2 + A_o^2 \right)^{0.5}}$$

in which A_i is the area of the inlet and A_o is the area of the outlet.

Assuming $A_i = 1.6$ m^2 for the door in our case, and A_o is the same for the roof light, then

$$Q = 0.93 \text{ m}^3/\text{s}.$$

If the volume of the house in Figure 9.15 is 200 m^3, the hourly air change rate would be about 17. Try comparing the above calculations with the ventilation of your own home in summer.

Note that the stack effect is proportional to the square root of the temperature difference and that the temperature difference is greater in the winter than in the summer. This is unfortunate in a sense because the driving force is greatest when it is least needed and least when it is most needed.

In the design of office buildings the forces acting on the building are, of course, similar to those on a house. As the loadings due to solar gains, occupants and equipment and the form of construction will, however, vary enormously, it is difficult to generalize, but as a very rough guideline single-sided ventilation is thought to be useful up to about 6 m.[24] Cross ventilation is thought to be effective up to above five times the height of the space; thus, for a 3 m floor-to-ceiling height the window-to-window distance might be 15 m.[25] In the author's offices – which are essentially circular in form (with a diameter of 23 m), have an average height of about 3.5 m and are open plan – cross ventilation from 22 windows located around the perimeter is adequate.

The BRE[26] has suggested that single-sided ventilation is effective up to about 10 m, based on tests of a south-facing room 10 m deep, 6 m wide and 3 m high; about 18% of the south façade was openable glazing. These findings require corroboration in practical studies and, for the present, it is probably best to consider 6 m as a maximum.

Where single-sided ventilation is not sufficient (or where noise limits its use) and where cross ventilation is not possible due to, say, compartmentation of spaces, stack effect ventilation may be a possibility.

Detailed knowledge of the effectiveness of these systems is limited but guidelines are being developed on the basis of theoretical studies and practical experience.[27,28,29] For example, a design guide for naturally ventilated courtrooms[30] recommends the following minimum figures:

courtroom height	= 6 m
stack height	= 5 m (*NB*: in this case the outlet air position could be below ceiling level)
free area of low-level inlet	= 1% of floor area
free area of high-level outlets	= 2% of floor area

At the De Montfort University Queens Building auditoria (Chapter 12) the as-built comparable figures are:

auditorium average height	= 5.1 m
stack height	= 12.0 m
free area of low-level inlets	= 4.8% of floor area
free area of high-level outlets	= 4.8% of floor area

Work is ongoing in the field and gradually knowledge of naturally ventilated buildings and ones with minimal mechanical energy is increasing.

Figure 9.17 summarizes some rules of thumb and design aspects that may prove useful for preliminary designs.

The procedure that is followed in developing real designs is, of course, more complicated. Essentially what is done is to assume an acceptable peak temperature in the space based on comfort considerations and discussions with the client, which thus gives the maximum temperature differential available to drive the stack effect. (This might be 2 °C above the ambient temperature.) Openings are then sized to give a ventilation rate. The ventilation rate is then put into a peak summer-time temperature calculation (using the admittance procedure, which takes into account the heat loads in the space and numerous other factors)[31] to see if the initial peak temperature and opening areas are suitable. The procedure is carried out iteratively until agreement is reached. Sophisticated computer modelling is also commonly used.

To transform Figure 9.17 into architecture is a challenge, and Table 9.2 lists some of the considerations; unfortunately, lack of space precludes anything other than a brief mention of one or two of them.

At the air inlet some kind of louvre may be used to allow air in and keep birds out while not causing too high a pressure drop. Figure 9.18 shows typical louvres with a free area of about 60%.

Louvres similar to those in Figure 9.18 have been used for some time to allow air entry to boiler rooms, as air inlets to mechanical ventilation systems and in acoustic enclosures. For natural ventilation systems their pressure drop must be examined closely.

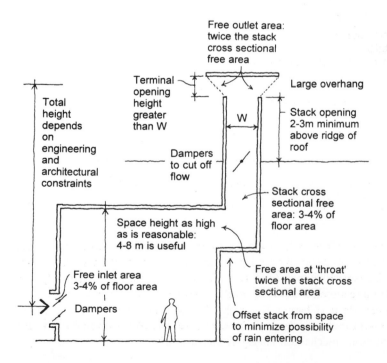

9.17 Simple stack effect ventilation.

Table 9.2 Checklist for naturally ventilated spaces

1. *Construction and configuration of space*

2. *Ventilation and expected internal temperatures*
    A. Stack effect
    B. Wind assistance
    C. Mechanical assistance
    D. Comfort and avoidance of overheating
    E. Room air movement

3. *Supply air*
    A. Source
    B. Tempering and heating generally
    C. Bird screens

4. *Extract air*
    A. Normal path
    B. Possibility of heat recovery
    C. Bird screens

5. *Smoke ventilation and fire prevention*

6. *Avoidance of rain and wind-driven snow*
    A. Prevention
    B. Draining if prevention fails

7. *Lighting*
    A. Daylighting and blackout facilities
        Interference with ventilation path
    B. Artificial

8. *Acoustics*
    A. Acceptable noise levels
    B. Room acoustics

9. *Risk of condensation*
    A. Interstitial
    B. On any part of exposed structure

10. *Controls*

11. *Durability, access and maintenance*

12. *Appearance*

13. *Cost*

Notes:
1. Naturally ventilated systems are not easily compatible with provision of incoming air filtration because of the additional resistance that results.
2. Wind loads need to be assessed by a structural engineer.

As we have seen in Chapter 4, acoustics must also be treated with great care. Mechanical ventilation systems owe their past success as much to their ability to deal with external noise as with temperature. Acoustic louvres on their own may only be a starting point. If more attenuation is required it will be necessary to introduce a path with a low pressure drop. (Chapter 12 briefly discusses how this was dealt with at De Montfort University. As can be seen there, the removal of noise usually requires space!)

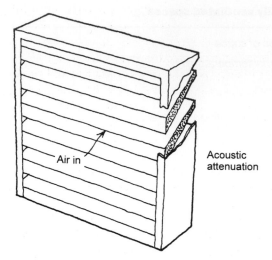

9.18 Louvre design.

Behind the louvres will be dampers that will probably be automatically controlled to shut off the air flow when it is not desired, such as when the space is unoccupied at night in winter. Figure 9.19 shows a detail of dampers with a free area of about 90% when open, and metallic side seals and blades tipped with rubber to minimize air flow through them when closed. When open the dampers must not, of course, have a high resistance to air flow.

Behind the dampers at some point there must be a heat emitter to ensure that before the air enters the occupied zone in winter it is 'tempered', i.e. brought up to a reasonable temperature that will avoid the discomfort associated with cold air movement. As with previous components, the heat emitter must not have a high resistance to air flow. The issue of draughts and air movement is important. If air is delivered under seats or close to people, possible discomfort must be considered. One approach is to deliver air at high level and let it mix with room air before it reaches the occupied zone. This can have the advantage of not causing high room air speeds at desk level in

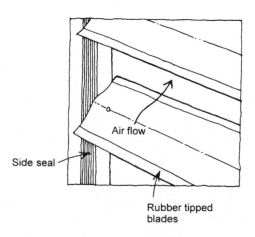

9.19 Damper detail.

offices, as air movements of 0.5 m/s can move papers. The situation requires careful examination, however, since, with some room geometries and ventilation patterns, the introduction of air at high level could displace a layer of hot stratified air downwards towards the occupants. In this respect, specialist advice may be needed.

In situations where noise is not an issue, the combination of louvre and damper can be dealt with by an open window that can be manually or automatically controlled. Windows must meet a variety of, at times, conflicting requirements in natural ventilation systems, and their design is not easy. They need to be easily openable and easily secured and any associated blinds or shades must also be easily operable.

Table 9.3 provides some thoughts on window types. Note that windows may also be used as the air outlets. Obviously some of the considerations listed apply to single-sided and cross ventilation systems as well.

The air outlets form an area where designs will be refined over the next few years. If we first consider areas, the free outlet of the stack should be about the same as the free inlet area, judging by studies for De Montfort University's Queens Building. The free area of the termination (sometimes called the monitor) should be about twice the free area of the stack. Again, the guiding rule is always to minimize the resistance to air flow. It is very desirable to have the termination open on all sides so that when it is windy, winds from all directions can assist the stack effect. For the same reason it is important to raise the termination above the ridge or any obstruction to remove it from the more turbulent zone created by wind passing over a building.

A wide variety of designs for tops of stacks can be made to work and so there is considerable scope for architectural expression. Solutions vary from the maltings type of Figure 9.11 to those of De Montfort (Chapter 12) to the lantern of St John's College, Cambridge, shown in Figure 9.20, which both admits light and exhausts air.

At St John's the areas ventilated by the lantern have doors that are held open magnetically and close on a fire signal. Generally, a great deal of attention needs to be given to smoke ventilation in assisted naturally ventilated buildings. Escape routes require careful consideration. There is also an alternative approach to that of St John's whereby buildings are compartmentalized and each compartment has its own ventilation system and fire control strategy (for example, see Figure 9.26).

It may prove of some benefit to make two final points on design:

1. Constant attention to detail is required when developing natural ventilation systems. The clash of a rainwater downpipe or a structural support with the air path, or spacing behind seats which creates draughts or inadequate controls and so on, are more likely to jeopardize success than with mechanical solutions.
2. On a less serious note, our rule of thumb is that the resistance of the air path is directly proportional to the number of designers and the length of the design process.

To conclude this discussion it should be said that more work is required in this field in order to optimize the type of system described above.

Assisted natural ventilation systems

If, moving towards the right of Figure 9.1, additional ventilation is required, it may be possible to add a simple fan to the system. This might take the form of a three-bladed propeller (or punkah fan), as shown in Figure 9.21.

Performance data for these fans is not always readily available. Our measurements for a typical 1350 mm diameter fan installed in a spacious vertical duct of low resistance (and so approaching 'free space') indicated a flow rate of about $0.7 \text{ m}^3/\text{s}$ when the fan was running quietly at about 80 rev/min.

Assisted natural systems have the great advantages of increasing the designer's (and client's) confidence that the system will cope with peak conditions. They also allow greater control and can be particularly advantageous for night-time cooling. Research is needed on the optimal mix of natural ventilation paths (which, of course, take up space and require construction materials) and simple mechanical fans. There is nothing sacred about 'pure' natural ventilation, and assisted natural ventilation systems are likely to provide the best way forward.

In theory it is possible to incorporate heat recovery in these systems because of the additional driving force provided by the fan, but in practice it may be more difficult. The resistance of the fan in the airstream may be excessive, the pressure loss at the heat exchanger can be too great and the electrical energy requirement may outweigh the energy recovered (similarly, the CO_2 balance may be unfavourable).

Typical heat recovery devices are shown in Figure 9.22. In cross-flow heat exchangers one stream of air transfers its heat to another across a series of plates. Thermal wheels rotate in two streams of air; the medium of the wheel picks up heat in the hot stream and transfers it to the cool one. Run-around coils contain a liquid (often water) that is used to transfer heat. Heat pipes are sealed tubes containing refrigerant. If one end is placed in a hot area the refrigerant evaporates and takes the heat towards the cold end of the tube where it condenses; the refrigerant returns to the hot end by gravity and capillary action in the wick. Reference 34 provides a useful introduction to these and other methods of recovering heat. All have been developed for use in mechanical ventilation systems and all tend to have high resistances (in part to increase their efficiency) in an approximate range of 150–300 Pa for the

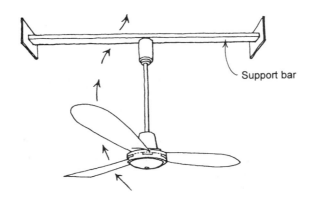

9.21 Propeller fan.

high velocities in mechanical ventilation systems.[36] Resistances are, however, likely to be lower because of the lower velocities in assisted natural ventilation systems. (In a recent project by the authors for a lecture theatre, run-around coils with a pressure drop of about 30 Pa and fans with duties of 1.6 m^3/s at 50 Pa were incorporated.)

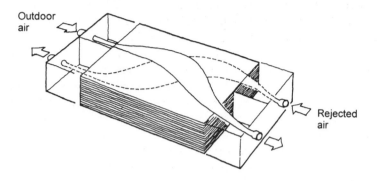

(a) Cross-flow heat exchanger

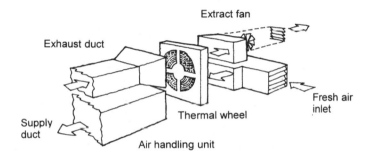

(b) Thermal wheel

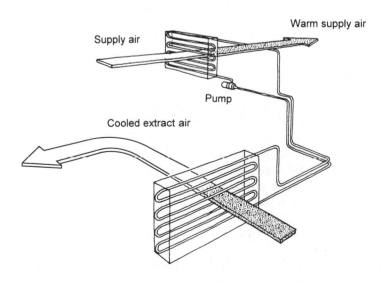

(c) Run-around coil

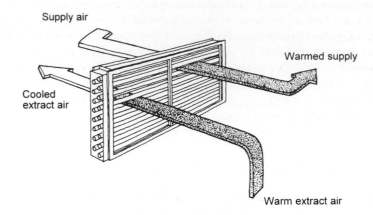

Supply air

Warmed supply

Cooled
extract air

(d) Heat pipe heat
exchanger

Warm extract air

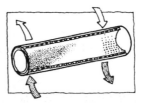

Detail of heat pipe

9.22 Heat recovery
devices.[35]

Approximate maximum temperature efficiencies (i.e. the difference between the temperature of extract from the space and the exhaust from the heat exchanger divided by the temperature difference between the supply air to the space and the incoming air to the exchanger) are 75% for cross-flow heat exchangers and heat pipes, 85% for thermal wheels and 65% for run-around coils.[37] System efficiencies are lowered by the extra energy required to force air through the heat recovery device. A fuller discussion on this topic can be found in reference 38.

One of the problems with most natural and hybrid systems is that the outlet and inlet airstreams tend not to be in proximity, making it more difficult to transfer heat from one to the other. Run-around coils are less affected by this problem and it may be economically possible to incorporate coils with wide spacings and, thus, lower pressure drops. Velocities will also be lower in hybrid systems and although this, too, will lower the pressure drop it will also decrease the heat transfer efficiency.

Where the inlet and exhaust can be brought closer together, or where the heat recovered might serve another space, the other alternatives could be considered. There may be some scope for retractable systems - for example, heat pipes that are only used during the winter and in summer are pulled out of the stack.

A degree of filtration may also be possible where fans are incorporated. Filters present very high resistances to air flows, and although they can function readily in mechanical ventilation systems there is almost no scope for their use in natural ventilation systems. Typical low-efficiency flat panel filters in mechanical systems (where duct velocities tend to be about 5 or 6 m/s) have initial resistance varying from, say, 25 to 250 Pa. As we have seen, this is much higher than the driving force of the stack effect and so generally, with natural ventilation systems, what is outside is what you will have inside (at least at the beginning). However, there is scope for coarse, lower resistance filters being used in conjunction with less powerful fans in assisted natural ventilation systems.

A general point in looking at natural, assisted natural and hybrid ventilation systems is that it makes sense to use existing low-resistance routes such as floor and ceiling voids, corridors and staircases. The key concern is that this needs to work in conjunction with the smoke control strategy.

9.6 Mechanical ventilation systems

Considerations of air quality and noise or stringent temperature and relative humidity requirements may necessitate standard mechanical ventilation systems, and these are the subject of a vast literature. Suffice it to say here that a ductwork system can be used

- in connection with central air-handling units, which may be just a fan but are more likely to incorporate a heater battery and may also have a cooling coil
- in connection with localized sources of a filter and heating, cooling or both.

One aspect that concerns us is the limitation of mechanical ventilation to deal with heat gains. Fans that bring in external air in the summer cannot lower the temperature of that air (indeed, because some of the energy needed to run the fans is normally transferred to the airstream, they will raise the temperature slightly). Thus, if a space is at 29 °C because of occupants, computers, lights and so forth, bringing in air at 25 °C will, at best, bring the temperature of the space down to 25 °C. In practice it is more likely that the temperature would only be brought down to 26 or 27 °C with the sizes of fans and air volumes likely to be used. And, of course, running the fans requires electricity and so results in CO_2 production (although in the nineteenth century, at least one bicycle-powered ventilation system was used in a London office building).

Mechanical ventilation with ductwork systems may be particularly advantageous in winter conditions when heat recovery is incorporated using devices such as those shown in Figure 9.22 and other equipment. Essentially there are two conceptual models, as shown in Figure 9.23. In model (a), which is the traditional type of building, air comes in and goes out via a variety of paths. The current tendency is to control these paths by more attention to construction details and by including such items as trickle ventilators. The great disadvantage of this model is that heated air is simply lost to the outside. In model (b) even more attention is given to control of the ventilation paths and great care is taken, for example, to seal the building, to incorporate high-performance windows which have very low leakage rates and to pressure test at completion to ensure that the construction is tight. It then becomes possible to introduce a controlled amount of air and extract a similar amount after recovering some of the heat it contains. This might seem to be an ideal system but it can be an expensive one. The additional capital cost of the fans, ducts and heat exchangers has meant that it has not been economically viable (given low energy

(a) Straight through natural ventilation – no heat recovery

(b) Mechanical ventilation with heat recovery

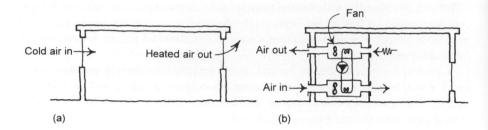

9.23 Ventilation models for the heating season.

costs). This is under continual review but the outcome is by no means clear. Let us consider the simplest situation of a gas-fired heating system being used in both models (a) and (b). With the fans of the latter being run electrically (as they almost always are), the current cost of electricity is very roughly three to four times that of gas, and so effectively 'expensive' electricity (and a variety of equipment that must be paid for) allows one to recover 'heat' produced from cheap gas. The amount of heat that can be recovered depends on the installation. For each project an economic analysis needs to be made and this must include some assessment of the relative costs of gas and electricity.

A more recent concern has been total CO_2 production for these two models. As we have seen, electricity produces much more CO_2 than gas. With an inefficient heat recovery system the CO_2 production can actually be higher than in a straight-through ventilated building.

What is clearly needed is comparative research into the actual performance of some buildings of both types. One example of model (a) can be seen in De Montfort University's Queens Building (Chapter 12). Model (b) is represented by the student residences at the University of East Anglia[39] and also by student housing incorporating an internal gallery at Christ's College, Cambridge. Figure 9.24 shows a simplified section through the building and a simplified schematic of the latter's ventilation system.

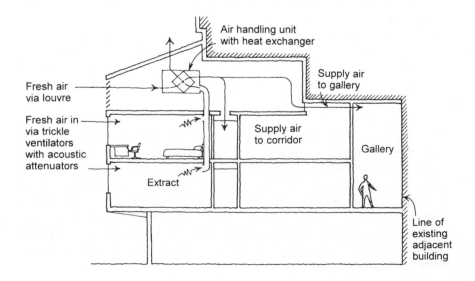

9.24 Ventilation with heat recovery at Christ's College, Cambridge, new student housing. (Architects: Architects Design Partnership.)

What factors tend to make it worth while to use a ducted air heating system with heat recovery? The first is, of course, availability of heat – a steady, ample supply of heat from people, equipment, lighting and, less reliably, solar gains is required. A fairly steady and significant demand is also, of course, required. A large number of people with the concomitant high fresh air requirement also helps because it means that the ventilation heat loss requirement will be significant compared to the fabric heat loss. Available space for the ductwork, is, of course, a prerequisite. And the more expensive the energy, the more worthwhile it is to recover it.

Developments in mechanical systems include displacement ventilation. Most mechanical (and natural) ventilation systems mix fresh air with indoor air to alter the temperature of the room air and reduce its level of pollutants. Usually this means that some of the pollutants will be breathed in. However, it is possible to direct air at people as, for example, in displacement ventilation where cool fresh air is introduced at low level and flows across the floor without a great deal of mixing. This pool of fresh air is pulled up by the convection currents generated by body heat (and computer and other low-level heat sources, if present). The polluted, warm air rises to high level above the occupied zone and is extracted. Obviously, the floor should be clean and the cool air should not be so cool as to make the occupants' legs cold.

9.7   Ventilation with cooling

If we now move to the extreme right of Figure 9.1, with the aim of keeping the internal temperature no higher (more or less) than ambient temperature, at some point it becomes worth while to consider the addition of a cooling coil to the mechanical ventilation system. If the aim is to keep the internal temperature below ambient temperature, a cooling coil (or some similar device) is almost obligatory. (The exception to this would be if the heat gains were low to moderate, ventilation requirements low and if the thermal mass of the building were high. The Pyramids and King's College Chapel use no cooling.)

For the past 60 years or so where cooling has been provided, for example in many office buildings around the world, it has tended to be based on CFC and, more recently, HCFC refrigerants. As we have seen in Chapters 2 and 6, most of these lead to depletion of the ozone layer and so CFCs have been phased out and HCFCs are following. At present alternatives include:

1.   Absorption cycle air-conditioning which uses, for example, ammonia as the refrigerant. This type of air-conditioning involves the absorption of the re-frigerant vapour in another substance, hence its name. (The more common cycle is the vapour compression type that is used, for example, in most buildings and at home in refrigerators and freezers.) The absorption unit may have gas as its basic energy source, in which case it, of course, produces CO_2. This argues for a building design that avoids the need for cooling, if at all possible. More sophisti-cated and often large-scale systems can link combined heat and power plants to absorption cycle AC units.

2. Ground water as a source of cooling (see Chapter 14 for a typical example and references 40 and 41 for others).

3. Vapour compression air-conditioning systems using HFCs (which have zero ozone depletion potential as shown previously in Table 6.1, although they do have a significant global warming potential) or other substances. These refrigerants should be regarded only as part of a transition to zero global warming, zero ozone-depletion solutions (See Chapter 6).

9.8 Controls

Controls are required to provide a comfortable environment with an economical use of resources – particularly energy. In the past this has been strongly influenced by costs and, with energy being cheap, relatively little was spent on controls. Now a combination of environmental problems and rising energy costs is forcing designers to give more consideration to controls.

There are a number of points that apply generally to controls. Firstly, the mechanical and electrical systems of a building and their associated controls should be as simple as possible, consistent with the need to meet requirements of efficiency, comfort and cost. A building that does more of the work will reduce the need for elaborate mechanical and electrical systems.

The more tightly something needs to be controlled, the more complicated, expensive and prone to breakdown the control system is likely to be. It is normally sufficient to let internal temperatures vary somewhat according to external conditions. This looser approach can minimize problems and save energy but it does require careful consultation among architect, engineer and client.

Try to imagine how the building will be used. Individuals want some control over their environments and need to be able to adjust them to avoid solar glare, or draughts, or uncomfortable temperatures. The needs of the individuals may conflict with those of the maintenance staff or with the operation of a central control system, and in most cases individuals should be able to override the automatic systems. In the future we are likely to see computer systems that allow such overrides but inform the individual if the action taken is not sensible.

Controls should be used to ensure that energy is produced only when it is needed and delivered only to the areas where it is needed. An important aspect of this is dividing the building into zones and ensuring that each zone is controlled suitably (but remember, too many zones are complex and costly). Zones can be thermal and so related to solar and internal heat gains, occupancy times, thermal response of the structure and type of heat emitter. Or they may be based on lighting, acoustic, smoke ventilation or other considerations.

Select control equipment that can be easily operated and locate it in an appropriate position. Controls must be comprehensible to those who are going to use them. As we become more computer-literate this poses less of a problem, but for the moment designers should speak with the people who will use the controls they specify to ensure that the controls match the people.

Controls should work with natural forces. One of the important developments in environmentally responsible buildings is to use controls to actuate lightweight elements that regulate internal and external forces. An example of this is a control system that operates lightweight dampers to let natural air flow through a space according to need (or demand). This 'intelligent' use of technology is an advance on the Jurassic forms of large air-handling units distributing significant volumes of air via extensive ductwork systems.

If we now look more specifically at controls for our services systems we can discuss them broadly by category.

Heating

Starting with heating, thermostatic radiator valves (TRVs), as shown in Figure 9.25, when fitted to heat emitters such as radiators or natural convectors will save energy provided, of course, that they are used. These valves work by restricting the flow of water to the emitter as the room temperature increases. An alternative that is somewhat more expensive is a thermostat in a space. This works in conjunction with an electrically operated valve (usually on–off) so that when the room temperature reaches the set point, flow to the heat emitters is shut off. Thermostats can be linked to time switches, and Figure 9.2 has shown a basic version of such an arrangement. Thermostats or temperature sensors linked to electrically operated valves can, of course, be used to control the entire range of heating emitters used in hydronic systems.

If we examine the operations of the boiler plant and heating circuits and move towards larger installations, energy can be saved by:

– including an optimizer in the system
– compensating the flow to each heating zone.

Optimizers switch the heating system on (or off) at a variable time that depends on outdoor and internal temperatures. They often incorporate 'learning' devices that can sense how the building responds. Thus, they are able to start the boilers earlier on a

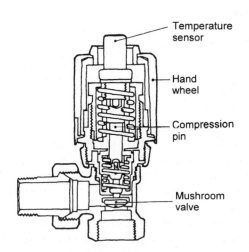

— Temperature sensor

— Hand wheel

— Compression pin

— Mushroom valve

9.25 Thermostatic radiator valve.[42]

cold morning than on a warm one. This ensures that the building will be comfortable by the required time and that only the energy needed to achieve this is used.

Compensating the flow refers to lowering the flow temperature at higher external temperatures. This improves energy efficiency by reducing unnecessary heat losses from pipework.

For larger systems numerous additional control strategies are available but they are too specialized to be covered here.

Ventilation – natural and assisted natural

In their simplest and traditional form, natural ventilation controls consist of manual opening and closing of windows and in some cases, such as nineteenth-century schools, ventilating panels built into walls and roof cupolas with opening flaps. A more sophisticated ventilation system, based on the auditoria at De Montfort's Queens Building (shown schematically in Figure 9.26), requires more elaborate controls.

Following the air path and starting at the inlet point, if there is any possibility of rain getting in dampers should close (either partially or completely) on a rain signal. The inlet air velocity may need to be controlled if high wind pressures can result in discomfort or papers blowing in the space, and this can be effected by a wind sensor. By incorporating a noise sensor it is also possible to ensure that peak noise is reduced.

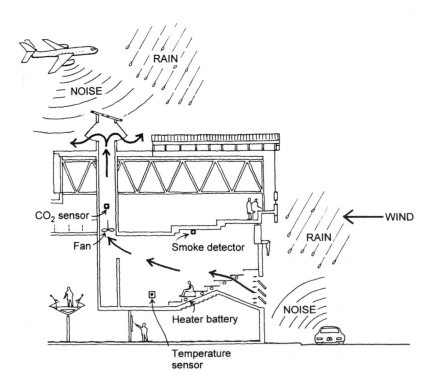

9.26 Automatic control strategies.

Incoming air needs to be tempered in the winter, and this can be done by control-ling the flow to the heater battery by sensors close to the occupants. Once the air has entered the space it will affect the occupants and be affected by them. It is important to incorporate a high degree of user control and it is recommended that manual overrides be provided. To determine the amount of air that should flow through a space to provide adequate fresh air and avoid stuffiness, a CO_2 sensor can be used.

Temperature sensors in the space can determine whether the open area should be increased and whether a fan should be brought on to assist the natural ventilation.

Smoke or heat detectors may be used to control dampers and windows. Early contact with the fire officer is essential in developing a successful approach to venti-lation and fire and smoke control. A common strategy, by which the space can be vacated quickly, is for high level exhausts to open and low level openings to close on a fire signal – the positions can be altered when the fire brigade arrives and determines the best strategy.

The control systems may incorporate rain and wind sensors at roof levels. These can close openings as required to ensure that rain does not enter the building envelope and that excessive air movement in the space that might result from wind-driven ventilation is avoided.

An important benefit of such a control system is that it can incorporate night-time cooling, thus, the temperature of the space can be lowered to some optimum value (related to its thermal mass and admittance) of, say, 15 or 16 °C by opening dampers and running the fan if necessary.

From the above discussion it is apparent that assisted natural ventilation systems like these are technically very sophisticated, even though they use much less traditional hardware such as ducts and air-handling units. With such systems natural ventilation can overcome one of its previous greatest disadvantages: lack of controllability.

Mechanical ventilation and air-conditioning controls

As with the systems themselves a vast specialized literature exists on their controls, and the interested reader should consult CIBSE and ASHRAE publications.[43,44]

Building management systems

Microcomputer-based building management control systems are key elements in the design of environmentally friendly buildings, especially larger ones. The principle of delivering energy only where and when it is needed assumes a network of sensors and a corresponding network of actuators such as switches and starters to run equipment or direct flows. Computers are, of course, well suited to this type of activity.

A typical layout for a building management system is shown in Figure 9.27. The system may vary in complexity but comprehensible simpler systems have obvious advantages in user-friendliness and cost. Systems whose computer routines are accessible and alterable by designers and users have definite attractions.

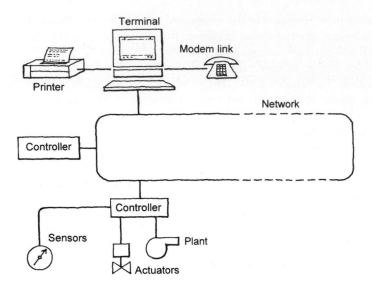

9.27 Typical configuration of a building management system.

Guidelines

Heating

1. Specify high-efficiency and condensing boilers.
2. Specify boilers with low NOx emissions.
3. The thermal response of the heating system must be matched to the building response and the pattern of occupancy.
4. Consider primary energy use and CO_2 emissions when selecting heating systems.

Ventilation

1. Use natural ventilation wherever possible, and avoid air-conditioning if you can.
2. Natural ventilation systems range from the simple to the complex. Do not underestimate the amount of thought required to make them work.
3. Assisted natural ventilation systems are likely to be appropriate, energy-efficient and reliable.
4. Give careful attention to noise and air quality.
5. Use heat recovery on extract air if you can prove it is economically viable and environmentally beneficial.

Ventilation with cooling

1. Cooling can be achieved without CFCs, HCFCs and HFCs.

Controls

1. Design the building to do as much of the work as possible. Mechanical and electrical systems and controls follow afterwards.
2. For larger projects building management systems are often an intelligent solution.

References

1. Lillywhite, M.S.T. and Trim, M.J.B. (1990). Gas-fired non-domestic condensing boilers. BRECSU Information Leaflet 18. BRE, Garston, p. 12.
2. Anon. (1990) *Guide for Installers of Condensing Boilers in Commercial Buildings*, Department of Energy, Efficiency Office Good Practice Guide 16, p. 2.
3. Anon. (2001) *ASHRAE Handbook – Fundamentals*, ASHRAE, Atlanta, p. 18.11.
4. www.sedbuk_com
5. Anon. (1988) *Condensing Boilers in Local Authority Family Housing*, Department of Energy, Energy Efficiency Office Expanded Project Profile 284, p. 2.
6. Anon. (2002) *CIBSE Guide B1: Heating*, CIBSE, London.
7. See reference 3, p. 15.15.
8. Anon. (n.d. *ca*. 1993) *Dunphy Combustion Environmental Technology Guide*, Dunphy Combustion, Rochdale.
9. See reference 3, p. 15.15.
10. Anon. (2004) BREEAM Offices 2004 Assessment Criteria. BRE, Garston.
11. Stephen, R. K. (1988) Domestic mechanical ventilation: guidelines for designers and installers. BRE Information Paper 18/88. BRE, Garston.
12. Anon. (2003) Good Practice Guide 345 Domestic Heating by Electricity. Energy Savings Trust.
13. Anon. (2003) *CIBSE Guide B4: Refrigeration and Heat Rejection*, CIBSE, London.
14. Banham, R. (1969) *The Architecture of the Well-Tempered Environment*, The Architectural Press, London.
15. Richards, J.M. (1958) *The Functional Tradition in Early Industrial Buildings*, The Architectural Press, London.
16. Ibid.
17. United Distillers, Edinburgh.
18. Walker, L. (1981) *American Shelter*, Overlook Press, Woodstock, New York.
19. Anon. (1999) *CIBSE Guide A4: Air Infiltration and Natural Ventilation*, CIBSE, London.
20. Anon. (1978) Principles of natural ventilation. BRE Digest 210. BRE, Garston.
21. Ibid.
22. Lacy, R.E. (1977) Climate and building in Britain, BRE, Garston.
23. See reference 19, pp. A4–5.
24. Walker, R.R. and White, M.E. (1991) Single-sided natural ventilation – how deep an office. BRE. *Air movement and ventilation control within buildings. 12th AIVC Conference, Ottawa, Canada.*
25. Anon. (1994) Natural ventilation in non-domestic buildings. BRE Digest 399. BRE, Garston.
26. See reference 24.
27. Martin, A.J. (1995) Control of natural ventilation. Technical Note TN 11/95. BSRIA, Bracknell.
28. Anon. (1997) Natural ventilation in non-domestic buildings. CIBSE, Balham.
29. Jackman, P.J. (1999) Air distribution in naturally ventilated offices. Technical Note TN 4/99. BSRIA, Bracknell.
30. Penz, F. (1990) *A Design Guide for Naturally Ventilated Courtrooms*, Cambridge Architectural Research Ltd, Cambridge.

31. Anon. (2002) *CIBSE Guide A5: Thermal Response*, CIBSE, London.
32. Baturin, V.V. (1972) *Fundamentals of Industrial Ventilation*, Pergamon, Oxford.
33. Anon. (1991) Code of practice for ventilation principles and designing for natural ventilation. BS 5925: 1991. British Standards Institution, London.
34. Hamilton, G. (1986) Selection of air-to-air heat recovery systems. BSRIA Technical Note TN 11/86. BSRIA, Bracknell, Berkshire.
35. Anon. (n.d.) *Heat Recovery with Heat Exchangers*, The Electricity Council, London.
36. See reference 34.
37. Anon. (1986) Heat recovery with heat exchangers. Electricity Council Publication EC3495/2.86. The Electricity Council, London.
38. See reference 34.
39. Evans, B. (1993) Airtight additions to the campus. *Architects' Journal*, **197**(17), 43–54.
40. Evans, B. (1994) Cooling and heat from groundwater. *Architects' Journal*, **199**(5), 23–5.
41. Bunn, R. (1998) Ground coupling explained. *Building Services* December, 22–7.
42. Anon (n.d.) *Taco-Constanta Thermostatic Radiator Valve*, Tacotherm, Basingstoke.
43. Anon. (2001) *Automatic Controls: CIBSE Applications Manual*, CIBSE, London.
44. Anon. (2004) *ASHRAE Handbook: HVAC Systems and Equipment*, ASHRAE, Atlanta.

Further reading

Anon (2000) Mixed mode ventilation. CIBSE AM 13. 2000, CIBSE, London.
Anon (2001) Ventilation and cooling option appraisal – a client's guide. Good Practice Guide 290. BRECSU, BRE, Garston.
Anon (2004) *CIBSE Guide F: Energy Efficiency in Buildings*, CIBSE, London.
Borland, S. (2004) Back to basics. *Building Services Journal*, 12/04, 43–4.
Kolokotroni, M. (1998) Night ventilation for cooling office buildings. BRE Information Paper 4/98. BRE, Garston.
Smith, P. (2003) *Sustainability at the Cutting Edge*, Architectural Press, Oxford.
Webb, B. and Kolokotroni, M. (1996) Night cooling a 1950's office. *Architects' Journal*, **203**(23), 54–5.
Welsh, P.A. (1995) Testing the performance of terminals for ventilation systems, chimneys and flues. BRE Information Paper 5/95. BRE, Garston.
Welsh, P.A. (1996) Flow resistance and wind performance of some common ventilation terminals. BRE Information Paper 6/95. BRE, Garston.

Water, waste disposal and appliances

10.1 Introduction

This chapter is a brief look at water, wastes and appliances. The one area examined in some detail is the provision of hot water services because of its importance in the design of larger, energy-efficient buildings.

10.2 Water

In the UK, availability of water is normally taken for granted and is only given any widespread attention in particularly dry summers with their resultant hose-pipe bans. But such an attitude is ultimately short sighted. Water is an essential natural resource whose collection, storage, treatment and distribution all have significant costs, for example schools in England spend approximately £61 million annually on water.[1]

There has thus been a slowly increasing acceptance of the need to conserve water by measures such as automatic flushing systems, self closing and spray taps, and of course, repairing leaking taps and pipework. Interest has turned to WCs with low water use given that around a third of domestic water use is used for WC flushing. The Water Supply (Water Fittings) Regulations 1999 now require a maximum flush volume not exceeding 6 litres.[2] Water consumption may be further improved by employing 'dual' flush volume devices.

Less conventional approaches to conserving water include the collection and storage of rainwater. Rainwater 'harvesting' systems have developed from systems designed in the 1970s[3]. Manufacturers now offer standard products, such as leaf filters, storage tanks, etc.[4] and rainwater collection is becoming an increasingly popular solution.

At Loughborough University swimming pool, rainwater is collected from the roof and stored in a tank below ground. The water is subsequently used to perform 'backwashing' of the pool filters which is a regular and water intensive process ($25m^3$ per filter).

10.3 Hot water service

Hot water demands vary according to building type and use. Energy consumption is also governed by the building layout and the system. Typical annual figures for energy consumption and CO_2 production for three types of building are given in Table 10.1.

Table 10.1 Approximate energy consumption and carbon dioxide emissions for hot water service systems[a]

Building type	Primary energy use for hot water service provision as approximate percentage of total energy use (%)	Delivered energy use (kWh/m² yr)		Carbon dioxide Production (kg/m² yr)	
		Typical	'Good'	Typical	'Good'
Office	1–15	20	2–5	4	2
School[b]	5–25	35	5–10	7	1–4
House[c]	15–25	40–50	20	10–30	4–6

[a] This is for a variety of systems and is based on an extensive review which included References 5–8. The figures shown reflect varying contributions from gas and electricity.
[b] Based on an approximation of one pupil per 8 m².
[c] Based on four people in a 100 m² house.

Energy use and, consequently, carbon dioxide production are governed by a number of factors, including:

— water use
— system considerations such as
    — choice of fuel and associated carbon intensity
    — concentration of demand
    — centralized vs decentralized systems
    — efficiency of water heating and control systems.

Water use will depend on the type of building, equipment installed and management techniques (to try to control use). A common starting point for analysis is to develop a demand histogram, as shown in Figure 10.1, which indicates the daily usage pattern.

Usage patterns can vary enormously: student residences have early morning and evening peaks while schools often experience a maximum demand at lunchtime for catering. Obviously the demand pattern can vary throughout the year, and again this is related to building type; schools will show a seasonal pattern but in office buildings consumption will be steady. Standard methods exist for estimating demand in a

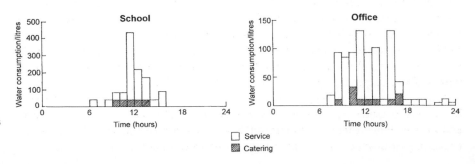

10.1 Demand histograms for a school and a commercial office.[9]

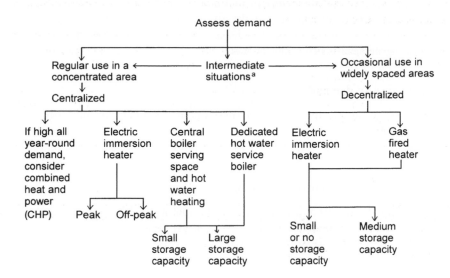

ᵃ Choice of systems for intermediate situations requires further details of each such situation.

10.2 Simplified selection of a domestic hot water system.

variety of building types,[10] and care should be taken to ensure that when applied they do not overestimate consumption for the building in question.

It is then important to evaluate the consequences of exceeding the likely demand; for instance, unavailability of hot water at home has very different consequences from unavailability in a hospital.

One of the most effective ways of reducing energy consumption is reducing hot water consumption. At the domestic scale an 80 litre bath uses about 48 litres of hot water each time compared to a five minute (9 ℓ/min) shower at about 27 litres. For larger buildings spray taps and flow restricters can reduce demand. Equipment – for example, clothes washers and dishwashers – that uses less hot water is obviously advantageous as demand reduction normally has follow-on benefits of lower storage capacities, smaller boiler and pipe sizes, and reduced quantities of waste water.

Choice of fuel

Presently in the UK the two main choices for fuel for hot water service systems are gas and electricity (see Chapter 7 for comments on other fuels). We have seen in Chapter 7 that, as a fuel, gas is both cheaper and has lower CO_2 emissions; however, the total system must be examined with the same precision as for heat recovery on extract air. Comparative studies for alternative systems in the same building are lacking, and studies of actual installations indicate that efficiency (defined as useful heat energy from the tap as a proportion of total energy delivered) ranges widely – figures for electric systems vary from 10–80% and for gas from 14–63%.[11,12]

Normally one might expect properly selected point-of-use electricity systems to be more efficient than gas (in terms of conversion of delivered energy to useful heat) and to have lower capital costs but perhaps higher running costs and greater CO_2 emissions. Each situation, however, must be analysed in detail.

It should be noted that the choice of fuel may be related to site considerations. If gas supply and gas-fired boiler(s) are already included in the engineering services design, the economics of gas and electricity is different from a situation in which no space heating demand is being met by gas.

Concentration of demand

In laying out the building the hot water draw-off points should be brought together as much as possible. This will result in both simpler and shorter pipework systems with less embodied energy, less pumping energy and reduced heat losses from pipework.

Centralized versus decentralized systems

At one end of the spectrum we have a concentrated area with a regular hot water demand served by a centralized hot water system. At the other end, we have widely spaced draw-off points with only occasional demands served by point-of-use heaters. If a centralized system were used in the second case, standing losses (e.g. heat energy conducted from a storage vessel or pipework) and perhaps circulating losses (e.g. electrical energy used by a pump to circulate hot water) would be too great and capital costs could also be significantly higher. One element of the latter is that hot water pipework must be provided whereas with point-of-use systems only a cold supply is required.

The point-of-use heater is likely to be electric if demand is small. If demand is greater, economics may favour a variety of gas-fired water heaters. There are architectural considerations in decentralization. Gas heaters of course need a gas supply and flue, and many architects eschew having a proliferation of flues. Similarly, if the system is 'unvented' (i.e. where the expansion volume is accommodated in a pressurized vessel) a safety valve and safe drain position are necessary and add complexity and a maintenance burden.

Maintenance is a significant factor in the choice. Some clients prefer to have their services centralized because they find it makes maintenance simpler and less costly. However, this depends on the expertise of the maintenance staff, and where the staff are less technically orientated, very simple decentralized systems may be easier to deal with.

There is scope for combining these two types of system. For instance, if most of a building works normal hours, but a small area uses hot water during the night, then the area with the 24-hour water demand could be served by a local heater, allowing the system serving the remainder of the building to be turned off at night. This approach has proved successful in hospitals.

If a centralized system is selected, is it better to use one boiler for both space heating and the hot water service or to separate the functions? Because space heating is seasonal and hot water service is much more constant there is an argument for a separate boiler for the hot water service. Such hot water boilers are available as packaged devices and offer the advantage of reduced energy loss in distribution pipework and pumping and dedicated controls. Each case, however, must be examined individually, and the economics assessed depending on the relative hot water and heating water demands.

The next consideration is likely to be boiler capacity versus storage capacity. Again, at one end of the spectrum we have a large boiler firing frequently and with no storage capacity and, at the other, a small boiler perhaps running almost continuously and with a large storage capacity, say, for 24 hours' demand. Most systems are, of course, between the two extremes and selection will depend on the demand profile and the space available for boiler and storage; to a much lesser extent it may depend on the peak fuel requirement and availability.

Boiler outputs should match the power required for the recovery time of the cylinders, and primary circuits should be pumped. This will allow reduction in the size of the pipes and thus will result in lower heat losses. Sequencing a series of boilers so that they operate near full load is common practice, but is often defeated by poorly configured controls.

The temperature of the water in the calorifiers and the distribution pipework is an important issue. Although the higher the temperature the greater the heat losses to the surroundings, common practice is now to store water at or above 55 °C, but provision for mixing of water from bath or shower taps must be made to avoid scalding. One reason for a stored temperature of 55 °C is to discourage the growth of Legionella bacteria which can occur at temperatures between 20 and 46 °C (37 °C is the optimum for multiplication).[13] Another is to avoid the waste of water that results when large quantities of tepid water are run off before hot water reaches the tap. At the calorifier a circulation pump can be used to mix the water, thus preventing stratification and reducing the risk of Legionella developing. The use of so-called 'destratification' pumps should be carefully considered when designing a system since stratification can also be used beneficially to reduce heat up time.

At the pipework there are two options: one is to have a hot water service (HWS) pump which returns water to the boiler as shown in Figure 10.3; the other is to trace heat the supply pipework with an electric tape under the insulation surrounding the pipe. The pumped solution requires a pump, of course, and return pipework but tends to be less expensive to run and to consume less primary energy. Again, there are no easy answers and the alternatives require economic and environmental analysis.

Combined heat and power

Figure 10.2 shows combined heat and power (CHP) as a possible source of heating. If there is a significant demand for electricity and a concurrent demand for a hot water service, CHP can be ideal. An illustration of this is that CHP is proving most cost effective in hospitals and hotels. Note that 'alternative' sources of energy, which include CHP, may in fact compete with each other. At the De Montfort Queens Building (Chapter 12), initial studies looked at both CHP and active solar collectors to meet the hot water service loads (and part of the space heating loads), and CHP appeared more economically viable.

Controls

Boiler controls should bring the boiler on only when needed and water should not be kept at a higher temperature than required. A combination of time clock control and a control thermostat on the calorifier should be sufficient. The controls will also pick up the boiler and the HWS and calorifier pumps.

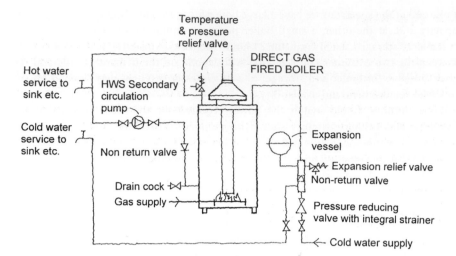

10.3 Hot water service system (simplified).

One of the most powerful aspects of controls is their ability to be adjusted to on-site conditions. This should be considered at every stage so that, for example, requirements can still be met efficiently if initial HWS estimates prove to be inaccurate in practice. Later, during use, being able to adjust the controls for the varying demand patterns that can occur will provide a more efficient installation.

Maintenance and management

An important but often overlooked aspect of efficiency is the maintenance and management of the installation. Those who manage the system should be familiar with how the designers intended it to function; they will then be in a better position to run it correctly initially and adapt it efficiently as conditions alter.

10.4 Waste disposal

Waste disposal systems that are not connected to mains sources include cesspools and septic tanks, and less conventional systems such as aerobic Clivus Multrum dry toilets and anaerobic methane digesters.[14] Cesspools are storage tanks which hold solids and liquids and need to be emptied regularly. Septic tanks retain the solid materials and discharge liquid to a means of secondary treatment such as a reed bed or below ground leach field before it enters the surrounding soil. A reed bed is a self-contained wetland ecosystem in which soil based microbiological processes promote the degradation of organic and chemical materials.[15] Clivus Multrum dry toilets are ventilated waste disposal units that take WC wastes and produce, after a number of years, a compost that has been found to be pathogen free.[16]

Methane digesters are sealed devices which convert waste materials into an effluent sludge and produce methane gas in the course of the decomposition process.

10.5 Appliances

Considerable scope exists for energy conservation (and, where applicable, reduced water consumption) with improved appliances. In the USA, for example, residential refrigeration consumes about 4.2% of the nation's electricity.[17] In Europe, household appliances for sale must display an energy label such as that in Figure 10.4. The label provides information on an appliance's energy performance as well as noise and water efficiency and consumption. An 'A' rated product is the most efficient and a 'G' rated, the least. Selecting 'A' rated appliances has been shown to be one of the most cost effective means of reducing CO_2 emissions from a building.[18]

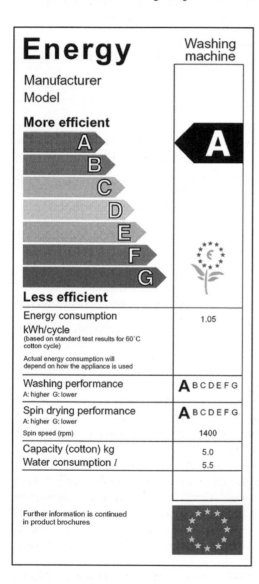

10.4 EU energy label.

Guidelines

1. Conserve water by the use of efficient distributors such as automatic flushing systems and low water content WCs.
2. Select the hot water service system that is appropriate to the building and the demand.
3. Alternative waste disposal systems are available and may be appropriate.
4. Select appliances that use environmentally friendly materials and have low energy consumption.

References

1. Anon. (2002) *Energy and Waste Management. A guide for schools*, DfES Publications. DfES/0209/2002. DfES, London.
2. Anon. (2002) Water Regulations Guide, Water Regulations Advisory Scheme.
3. Littler, J.G.F. and Thomas, R.B. (1984) *Design with Energy*, Cambridge University Press, Cambridge.
4. Anon. (2004) www.rainwaterharvesting.co.uk
5. Anon. (1998) Saving energy in schools: a guide for headteachers, governors, premises managers and school energy managers. Energy Efficiency Best Practice Programme: Energy Consumption Guide 73. BRE, Garston.
6. Anon. (2003) Energy use in offices. Energy Efficiency Best Practice Programme: Energy Consumption Guide 19. BRE, Garston.
7. Veley, J. I. et al. (2001) Domestic energy fact file: England, Scotland, Wales and Northern Ireland, BRE Report 427. BRE, Garston.
8. Pout, C. et al. (1998) Non domestic buildings energy fact file, BRE Report 339. BRE, Garston.
9. Anon (1999) *CIBSE Guide G: Public Health Engineering*, CIBSE, London.
10. Ibid.
11. Anon. (1986) *CIBSE Guide B4: Water Service Systems*, CIBSE, London.
12. Skegg, V.E. (1982) Losses in DHWS distribution systems. *Health Services Estate*, **48**, 46–53.
13. Anon. (2002) CIBSE Technical Memorandum 'TM13:2002. Minimising the risk of Legionnaires' disease. CIBSE, London.
14. See reference 3, pp. 293–303.
15. Grant, N. and Griggs, J. (2001) Reed beds for the treatment of domestic wastewater. BRE, Garston.
16. Anon. (2005) www.clivus.com
17. Anon. (2005) www.eia.doc.gov
18. Thomas, R. (2003) *Sustainable Urban Design – An Environmental Approach*, Spon Press, London.

Further reading

Anon. (2005) CIBSE Knowledge Series: Reclaimed Water. CIBSE, London. www.watermark.gov.uk

Summary

This section briefly summarizes the preceding chapters and introduces the case studies that follow.

An overview

From the general to the particular and back has been the underlying approach so far. The general laws of physics govern your immediate environment and the chair you are sitting on is related to the global ecology.

Design starts with the site and continues through a choice of form, a selection of materials and development of the engineering systems that provide what the natural environment and building cannot.

There is no clear-cut distinction between architects and engineers in an environmentally responsive architecture. Frank Lloyd Wright said that the development of piped heating systems allowed him to articulate the building form. In current design, in part because of the development of improved insulation materials and the often high internal loadings of buildings, we are led to articulate the building to exploit the natural forces of Sun and wind. To do so requires a common understanding on the part of architects, engineers and builders. The case studies which follow are examples of a cooperative approach to the design and realization of a number of exceptional buildings.

It will be clear from them that there are no blueprints for low-energy, environmentally friendly buildings but that there are a number of ways of achieving comfortable conditions efficiently in both summer and winter.

Finally, it is worth while emphasizing that we are only starting – almost everything about an ecological approach to our environment remains to be done.

An introduction

The case studies selected illustrate a number of the principles discussed so far. Our debt to their architects and the other members of their design teams is enormous. The studies should be regarded as work in progress – within the short period they cover energy efficiency has improved and new techniques have been tested and proved successful. Among the important points they make is that an environmental approach not only need not constrain architectural creativity but can encourage it and that environmentally sensitive buildings can be of the highest architectural quality.

Part Two

RMC International Headquarters CHAPTER 11

11.1 The site and the building

RMC House, completed in 1989, is the headquarters building for Ready Mix Concrete PLC, at Thorpe in Surrey. The site contained three existing houses and the architect's intention was to create a complex in which they were incorporated. Because the site is in the 'green belt' of London the architect strived to ensure that the headquarters were well integrated with the surrounds and that the public appearance was discreet. One approach was to design a largely single-storey structure with an extensive roof garden, as shown in Figure 11.1.

The building is about 10 000 m^2 in total, of which 4500 m^2 is offices. There are also training facilities in the form of various lecture rooms, an audiovisual studio, a laboratory, and an amenity complex with a swimming pool, two squash courts and a gym for the use of the staff. Finally, there is accommodation in the form of study bedrooms for the trainees.

One intention of the design was to keep the offices comfortably cool without air

11.1 Roof gardens.

conditioning. This involved the application of passive techniques to reduce the heat gains in the offices and to mitigate their effects and was achieved by:

1. providing internal and external solar shading;
2. insulating the building to a high standard;
3. incorporating significant mass in the building for cool thermal storage;
4. use of the ground floor slab as a cooling element for the incoming air in conjunction with a mechanical ventilation system;
5. providing numerous openable windows to allow cross ventilation;
6. increasing the daylight penetration into the building to reduce the need for artificial lighting.

The roof build-up is 500 mm of soil on top of 100 mm of polystyrene insulation, on top of asphalt on a 150 mm concrete slab. This gives a U-value of 0.3 W/m^2 K which results in a low heat loss through the roof. The great thermal mass of the roof construction also provides a time lag of about 12 hours, which delays the external heat reaching the offices. This means that the heat of the day only reaches the offices at night, and that during the day the roof slab radiates, in a sense, the cool of the previous night into the offices.

The building is divided up by courtyards which allow daylight to enter the offices. One of these courtyards is paved with white tiles to increase the amount of daylight into the adjacent spaces. The walls that face the courtyards have full-height double-glazed patio doors; internal walls are mostly lightweight plasterboard, or glazing. Some of the deep plan internal areas have glazed rooflights, as shown in Figure 11.2; Figure 11.3 shows a similar area in plan.

In Figure 11.4 the perspective drawing shows the construction and the results of a calculation of the room admittance. (The concrete slab under the floor tiles contributes significantly to the room admittance.) Typical room admittances are 8 W/m^2 (floor area) K for lightweight constructions and 24 W/m^2 (floor area) K for very

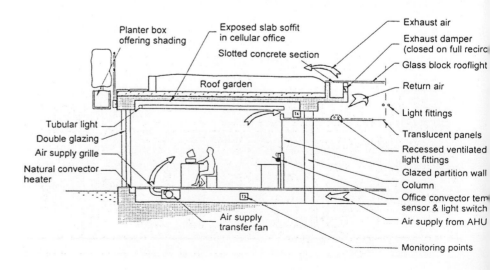

11.2 Section through a deep plan office area.

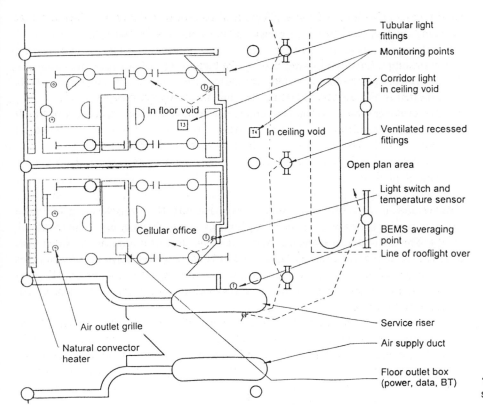

Tubular light fittings

Monitoring points

Corridor light in ceiling void

In floor void

In ceiling void

Ventilated recessed fittings

Open plan area

Light switch and temperature sensor

Cellular office

BEMS averaging point

Line of rooflight over

Air outlet grille

Natural convector heater

Service riser

Air supply duct

Floor outlet box (power, data, BT)

11.3 Office plan with services.

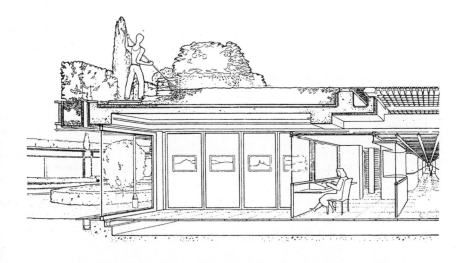

Medium weight room admittance 18 W/m² (floor area) K

Ceiling, concrete slab; internal walls, plasterboard; external walls, double glazing; floor, carpeted timber floor tile with concrete slab under

11.4 Typical office and room admittance. (Drawing by Edward Cullinan.)

heavyweight construction.[1] External shading is provided by planter boxes which overhang the roof edge. Patio doors have internal venetian blinds. The core areas of the building have plasterboard false ceilings, but the perimeter offices have exposed concrete soffits, thereby allowing the heavyweight roof slab to help moderate the office temperatures. In some corridor areas there are rooflights of translucent panels (as indicated in Figure 11.3) to reduce the need for artificial lighting.

The perimeter offices have natural convector heaters under the patio doors which offset the heat losses through the glazing. Hot water is provided by central gas-fired boilers.

The offices are ventilated either by individuals opening their patio doors, or by mechanical air-handling units (Figures 11.2, 11.3 and 11.5). There are a number of compact air-handling plant rooms on the roof of the building, each of which serves roughly 500 m^2 of office. The air-handling units (AHUs) supply air down false columns to the floor void, which then acts as the supply ducting throughout the offices. The air is drawn from the void and blown out to the offices by small fans which serve four swirl type air outlet grilles. Air is extracted from the perimeter offices at high level in the false ceiling area, which is over the central offices. This area acts as a route for this extract air back to the rooftop plant rooms. From the central false ceiling area the air is either returned to the air-handling plant rooms to be recirculated, or is exhausted to atmosphere through specially designed motorized

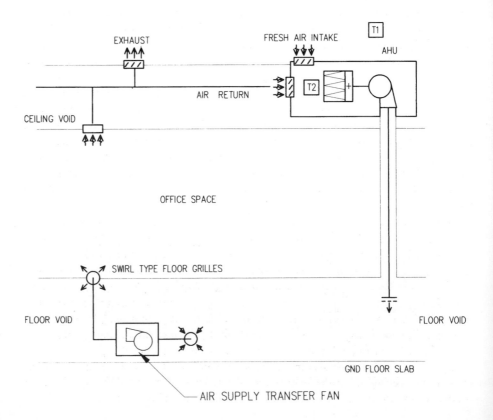

11.5 Schematic of mechanical ventilation systems.

dampers under the rooflights. The air from the central office areas is extracted through the light fittings into the false ceiling zone. This removal at source ensures that little of the heat that is generated by the lights reaches the occupants.

The air-handling units were sized to provide six air changes per hour to the offices or about 5 litres per m² floor area. This air flow was calculated to result in a maximum internal temperature of 25 °C when it reached 27 °C outside in conditions other than extended heatwaves.

The air-handling units generally recirculate the air, providing a small percentage of fresh air. The exceptions to this are:

–   during the heating-up period in the morning when the air is 100% recirculated
–   in summer when free cooling is possible, and in this case the air is 100% fresh.

Free cooling describes a situation when the offices are above their temperature set point and so require cooling, and the external air temperature is lower than the internal air so that cooling is possible. Cool air is taken from outside and passed over the ground floor slab under the raised floor. This is normally done at night by running the ventilation system and results in a process of cool thermal storage in the ground floor slab and the roof slab, which both have significant thermal mass. Perhaps the simplest way to think of what happens is to consider the slabs as storage coolers (rather than the storage heaters with which we are more familiar).

At night the fans will only run if there is a sufficient temperature difference between internal and external air to make their operation efficient and economic. If it is hotter outside than inside, the fans will not run as no cooling can be obtained. If it is only slightly cooler outside, then it is still not beneficial to run the fans as the heat gain to the air from the fan motors will heat the air to above the temperature of the internal air.

In some cases it is beneficial to encourage stratification of the temperatures in buildings in summer. Stratification is the situation whereby air at higher temperatures lies in layers above air at lower temperatures. Generally, but depending on the geometry of the space and ventilation rates, occupants will tend to be in the lower cooler areas. In this building this effect was intentionally reduced so that the roof slab could be kept as cool as possible in order for it to radiate to the occupants the following day. (In winter stratification is not desirable because the heat is needed at low level.)

The lighting in the cellular offices consists of suspended single-tube pipeline type fittings (Figure 11.2). The open plan offices have recessed, ventilated, twintube fluorescent fittings and bare single tube fittings between the translucent ceiling panels and the rooflights. The light level is about 400 lux in the central areas and about 600 lux in the cellular offices, with installed loads of 16 and 22 W/m² respectively. These figures are somewhat high because the lights provide some uplighting as well as downlighting, and, to a lesser extent, because high-frequency ballasts were not used.

The uplighting is less efficient because the reflectance of the ceiling is only about 60%, thus 40% of the uplighting energy input is lost immediately.

Lights in the cellular offices are individually switched. In the open plan areas lights are switched centrally, but the switches are always local to the areas served so that the occupants are encouraged to switch off lights that are not required.

Most of the occupants of the cellular offices are senior executives, with administrative staff in the open plan areas. The occupancy is one desk per 20 m^2 in the cellular offices, and one desk per 30 m^2 (including circulation space) in the open plan areas giving a low loading from occupants of roughly 4 W/m^2.

Generally, in the building there are centralized photocopying facilities. Together with distributed small machinery such as personal computers and coffee machines, the loads are again low at about 7 W/m^2.

11.2 Internal conditions

The temperature inside the building is very stable, with an amplitude (maximum to minimum) of less than 3 °C on a hot day as can be seen in Figure 11.6. This stability is greater than might have been expected and is attributable to the large area of high admittance surfaces both directly connected to the space, such as the (coffered) ceilings, and indirectly connected via the ventilation system, such as the floor slab. The temperature of the floor slab remains very constant at 20–21 °C.

Other monitoring has shown that when the external air is at 28 °C, the air from the floor grilles is 2.5 to 3 °C below that temperature. The fans themselves raise the temperature of the air about 1 °C so the cooling effect of the slab is 3.5 to 4 °C. Given that the temperature difference between the external air and the slab is roughly 7 °C (from about 28 °C to 21 °C) the heat transfer efficiency is very good.

Analysis of extensive measured data has shown that the minimum ground floor slab temperature lags behind the minimum external air temperature by about 10 hours. Further examination also showed that the slab can take 4–6 days to respond to a heat wave. Thus, it can be concluded that the ground slab functions effectively as a buffer or store.

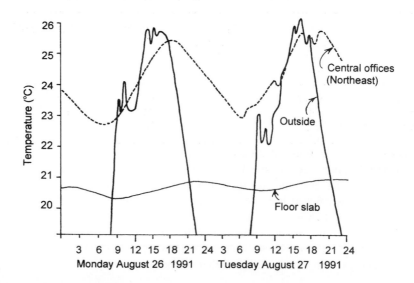

11.6 Temperature data for the central office area.

11.3 Energy consumption

Figures provided by RMC show that the building used 1484 MWh/yr of electricity and 2676 MWh/yr of gas on average in 1991 and 1992. Table 11.1 gives a comparison of the energy consumed in the RMC office area and 'typical' and 'good practice' office data from the Energy Efficiency Office;[2] the RMC figures have been estimated using data from the building management system and details of the installed loads.

The figures in Table 11.1 indicate the generally good performance of the building. Electrical energy consumption is somewhat high because of the lighting as discussed previously.

11.4 Conclusions

The offices at RMC House are performing well and are reasonably energy efficient. In summer where the occupants have access to patio doors they have control over their own environment which leads to a high degree of occupant satisfaction. There have been some complaints from the people in the central area of the building, mostly due to a feeling of stagnant air, and probably exacerbated by the lack of outside awareness and individual control.

The high thermal mass of the building in conjunction with the underfloor ventilation system produces a stable floor slab temperature and comfortable conditions in the offices without the need for air conditioning.

There is no spare capacity, available with the flick of a switch, in the performance of a passively cooled structure. Therefore it is important to fine tune a number of aspects of the building. This is best done by monitoring performance and this is easiest with a building management system. The interest and expertise of the management and maintenance staff have also been key elements in this headquarters' success.

Table 11.1 Comparative delivered energy consumption data (kWh/m^2 yr)

Building	Heating and hot water services[a]	Fans, pumps and controls[b]	Lighting[b]	Equipment[b]	Other[b]	Total electrical
RMC offices	114.2	17.2	47.3	13.5	4.2	82.2
Typical office	200	21	53	16	5	95
Good practice office	95	13	35	16	4	68

[a] Provided by gas.
[b] Provided by electricity

Project principals

Client	Ready Mix Concrete PLC
Architects	Edward Cullinan Architects
Services engineers	Max Fordham LLP
Structural engineers	Anthony Hunt Associates
Quantity surveyor	Raymond Hart & Partners
Landscape architects	Derek Lovejoy & Partners
Project managers	Pelmaks Project Management Services Ltd
Management contractors	Trafalgar House Construction Management
Mechanical contractors	Rosser & Russell
Electrical contractors	N.G. Bailey

References

1. Petherbridge, P., Milbank, N.O. and Harrington-Lynn, J. (1988) *Environmental Design Manual*, BRE, Garston.
2. Anon. (1991) Energy efficiency in offices. Department of Energy, Energy Efficiency Office Best Practice Programme: Energy Consumption Guide 19. HMSO, UK.

Further reading

Davey, P. (1989) Musique concrete. *Architectural Review*, **188**(1123), 59–67.

Queens Building, De Montfort University

The Queens Building at De Montfort University, shown in Figure 12.1, which houses the school of engineering and manufacture, is one of the most exciting and innovative buildings to have been built in Europe in the last few years. The building represents a major shift from the tradition of university buildings that has grown up over the past 20 to 30 years. The design strategy chosen has been to look to the solutions of the past then, applying modern technology, to refine these solutions and make them the way to the future. In particular, the building has been designed to function without the use of air conditioning in spite of high internal heat gains in lecture theatres and computer laboratories. The result at De Montfort is a building that is wholly modern with a strong bias towards low-energy design which, with its red brick construction, fits well into the traditional Leicester cityscape.

12.1 External view of the building.

Table 12.2 Maximum design internal heat gains

Space	Area (m²)	Solar gain (W)	Equipment gain (W)		Occupancy (100 W/person)	Total (W)	Total (W/m²)
Electrical laboratories	170	4500	Computers	6000	3000	15110	88
			Lighting	1610			
Central building Auditoria	150	–	Audio-visual	500	15000	17600	117
			Lighting	2100			
Mechanical laboratory Machine hall	644	33000	Equipment	10000	2500	70500	109
			Lighting	25000			

The north side consists of two 70-seat lecture theatres below the double-height 150-seat auditoria at first-floor level. At third-floor level the drawing/design studios are traditionally designed to have a combination of top and north light.

The two sides of the central building are separated by a long narrow central concourse space which rises through the full height of the building and is top lit. This acts as the central communication route at ground- and first-floor level between the electrical and mechanical laboratories. In addition, it is bridged at third-floor level by flying walkways which link the staff offices to the drawing studios. The drawing studios overlook and are open to the main concourse volume.

The mechanical laboratories are the third main area and are located at the west end of the building. The main space is a double-height machine hall with a control room on a mezzanine level at each end. Adjacent to the machine hall at ground- and first-floor levels are smaller rooms which are dedicated to specific mechanical engineering functions such as welding, printed circuit board production and metrology studies. Owing to the nature of the work carried out in these ancillary rooms and the presence of fume cupboards to handle chemicals, two small air-handling units have been installed to provide these rooms with mechanical ventilation (but not cooling).

12.2 Natural ventilation

One of the main design principles of the building was that it should be naturally ventilated because of the client's and design team's commitments to environmental issues.

The two auditoria proved to be the most heavily loaded areas from the point of view of internal heat gains, followed closely by the electrical engineering computer laboratories. Internal gains in the mechanical laboratories, and in particular, in the machine hall were more difficult to predict because although there are some very large items of equipment their use tends to be intermittent.

As we have seen previously (Chapter 9), in naturally ventilated buildings air moves because of pressure differences arising from stack effect and wind effects. Stack effect or buoyancy forces are caused by warm air rising and being replaced by cold air at low level. Wind effects arise due to wind flow around the building causing pressure gradients over the building envelope. Normally, both effects are in operation, but the contribution of each varies with the geometry of the space, the size and positions of air intakes and exhausts and the temperature differential between inside and outside.

The new engineering school has a range of different combinations of stack and cross ventilation. The electrical laboratories, because of their narrow plan, are cross ventilated and wholly manually controlled via opening windows. The window openings were sized using a low design wind speed of 0.5 m/s, which is the wind speed expected to be exceeded for more than 90% of the time. Table 12.3 shows the equations used in the calculations of the opening areas required for a typical cross-ventilated electrical laboratory using the heat gain figures given in Table 12.2.

The more deep plan spaces of the central building that are open to outside air on one face only are stack effect ventilated, using, in the case of the ground-floor classrooms and first-floor central laboratory, a combination of manual air intakes (opening windows) on the perimeter and automated air exhausts via stacks or openable roof-lights deeper into the plan.

The auditoria which, as we saw in the worked example above, are also stack effect ventilated gave rise to the dramatic stacks above the roof line. The brief from the client for the auditoria called for the seating of 150 people in a space that could be blacked out. The thermal and acoustic environments did not have to meet the stringent criteria of an air-conditioned auditorium. This is one of the central issues relating to providing low-energy and naturally ventilated buildings – the client must be prepared to accept a more natural environment with its associated changes in temperature in the summer as the day warms up and, indeed, changes in noise levels as traffic passes by outside.

The occupancy, together with solar gains and gains from lighting and equipment, made the auditoria a main focus of the design. It was agreed at a very early stage that the lighting would have to be fluorescent to keep the heat gains as low as possible compared with more traditional tungsten sources. This decision also means lower maintenance costs as the lamp life of the low-energy fluorescent fittings used is approximately 8000 hours compared to 1000 hours for a tungsten source. The architects included the auditoria in their initial computer analysis of stack effect ventilation. These preliminary studies suggested that a loading of 100 W/m^2, an air inlet area of 2.5 m^2 and a stack height of 7 m would result in a peak internal temperature of 27 °C, given an external peak of 24.5 °C.[1]

Subsequent theoretical testing included salt water modelling by Cambridge Architectural Research Limited in conjunction with the University of Cambridge Department of Applied Mathematics and Theoretical Physics. A perspex model of the auditorium was built and immersed upside down in a bath of water. Salt solution was then dyed and injected into the perspex model and the resultant water flow patterns were filmed and analysed. The density of the salt water could be increased or decreased to represent higher or lower heat gains, respectively. The filmed results of the tests were inverted so that the salt water dropping could be seen as warm air rising. The model showed that connecting more than one space into a stack could result in air flow from one room to the next rather than upwards to exhaust at the top of a stack. On this basis individual spaces were given dedicated stacks. This, in turn, coincided with the requirements of the local fire prevention authority for smoke

Table 12.3 Worked examples for calculating the opening areas required for natural ventilation using equations from Chapter 9

Stack effect – Path 3 (Figure 9.15) as applied to an auditorium

Under conditions where there is a 5 °C temperature difference between inside and outside and the height of the stack is 12 m

$$\Delta p = 0.043 \, h \, (t_{int} - t_{ext}) \text{Pa}$$

$$\Delta p = 0.043 \, (12) \, (5) = 2.58 \text{ Pa}$$

$$A = \frac{A_i A_o}{(A_i^2 + A_o^2)^{0.5}}$$

$A_i = 4.9$ m^2, i.e. the free area of the air intake from the road; $A_o = 5.2$ m^2, i.e. the free area of the air exhausts.

$$A = \frac{(4.9) \quad (5.2)}{(24.0 + 27.0)^{0.5}} = \frac{25.48}{7.14} = 3.6 \text{ m}^2$$

$$\begin{aligned} Q &= 0.827 A (\Delta p)^{0.5} \\ &= 0.827(3.6)(2.58)^{0.5} \\ &= (0.827)(3.6)(1.6) \\ &= 4.8 \text{ m}^3/\text{s} = 17\,280 \text{ m}^3/\text{h} \end{aligned}$$

The floor area of an auditorium is approximately 150 m^2 and the average height is 5.1 m, thus the volume is 765 m^3. This gives an air change rate of 23 per hour when there is a 5 °C temperature differential between inside and outside.

The fresh air requirement for 150 people in the auditorium is 10 l/s person × 150 = 1.5 m^3/s . This is significantly less than the air movement caused by the stack effect in this example, and in winter conditions the air intakes and exhausts can be opened only slightly to allow the ingress of enough fresh air for the people in the space. In winter the temperature difference between inside and outside is much greater than the 5 °C used in this example, and if the air intakes and exhausts were fully opened there would be a very high air flow through the space and a huge heat loss.

Wind effect – Path 2 (Figure 9.15) as applied to the top floor of the electrical laboratory

Cross ventilation by the wind alone is given by

$$Q = 0.61 \, A_w u (\Delta C_p)^{0.5}$$

where 0.61 is the discharge coefficient and A_w is the equivalent area (m^2) for wind-driven ventilation given by

$$\frac{1}{A_w^2} = \frac{1}{A_1^2} + \frac{1}{A_2^2}$$

In this case, assume ΔC_p to be 0.5 (ΔC_p is the change in the pressure coefficient; ΔC_p ranges from 0.1 for a sheltered site to 1.0 for an exposed site). u is the velocity, taken as 4.0 m/s, which is our assumed wind speed at a height of 10 m exceeded for 50% of the time (Appendix A).

$$A_1 = 2.535 \text{ m}^2 \text{ and } A_2 = 2.535 \text{ m}^2$$

$$\frac{1}{A_w^2} = \frac{1}{(2.535)^2} + \frac{1}{(2.535)^2} = 0.31$$

Table 12.3 (continued)

$\therefore A^2_w = 3.22$ and $A_w = 1.79\ m^2$

$Q = (0.61)(1.79)(4.0)(0.5)^{0.5} = 3.10\ m^3/s.$

Volume of electrical laboratory
$$= 170 \times 3.3$$
$$= 561\ m^3$$
$$3.10\ m^3/s \times 3600 = (11\,163\ m^3\ \text{of air/h})/561$$
$$= 20\ \text{air changes/h.}$$
To estimate the temperature rise in the space at the flow rate available, and taking the total heat gain in the electrical laboratory from Table 12.2 as 15.11 kW or 15.11 kJ/s.

$$\frac{15.11}{1.2 \times \Delta T} = 3.10\ m^3/s$$

(where 1.2 is the specific heat capacity of air measured in kJ/m³K).

Thus the temperature rise in the space ΔT is 4.1 °C. If the amount of air flow calculated in this way gives a temperature rise in the space which is considered unacceptable, then further openings should be introduced to increase the air flow which will, in turn, reduce the temperature rise. In practice, both the heat loads from computers and the temperature rises have been less.

---

control in a fire situation. The weakness of the perspex model is that it could not take account of admittance (i.e. the use of the building fabric as a thermal store).

In addition to the salt water model, the School of the Built Environment at De Montfort University carried out computer simulations using the ESP package. The authors carried out the standard Chartered Institute of Building Services Engineers admittance procedure and stack effect calculations. Table 12.3 shows simplified versions of the equations used for the stack effect calculations.

These calculations were done both as a check on the more sophisticated techniques mentioned above and because design liability lies with the design team. A figure of 10 litres of fresh air per second per person was used for odour control and we chose an external design day of 25 °C maximum, 19 °C average and 13 °C minimum. The client was informed and agreed with this choice of day. The results showed that, in general, peak internal temperatures would not rise by more than 4–5 °C above peak external. This was a conservative figure based on temperature difference only (i.e. stack effect calculations) and not wind effect. As a back-up for the occasional very still, hot day a punkah fan of duty 1.7 m³/s at 10 Pa was included in one of the two stacks in each of the two auditoria to provide some minimal mechanical assistance.

A comparison of the results from the various calculation techniques is shown in Table 12.4. Actual results, as shown in Figure 12.8, are in fact significantly better than predicted.

Figure 12.3 shows a section through one of the auditoria. Air enters at the street side through large openings protected from the weather, passes through motorized volume control dampers at the building envelope line and through an acoustically lined plenum and is distributed through the void under the seating. It passes over finned heating tubes suspended under the seats, out under the seats and then through a grille made of aluminium mesh. There are no air filters, which would create too high a resistance to air flow.

Table 12.4 Results of natural ventilation analyses for the auditoria

Technique	Peak internal temperature[a] (°C)	Comment
Saltwater modelling	29.8[2]	Stack effect only, thermal inertia not accounted for
Computer modelling	28.9[3]	Wind effects were simulated in addition to stack effect
Admittance procedure	29.6	Stack effect only; the technique is relatively crude and ventilation rates are rough estimates

[a] Temperatures have been adjusted relative to an external peak temperature of 25 °C (based on a day with a 13 °C minimum and 19 °C mean).

The air that is heated by the occupants and other internal gains, such as from lighting and audiovisual equipment, then rises and travels into the diamond-shaped exhaust stacks which also serve as supports for the drawing studio roof trusses. The total cross-sectional free area of the two stacks is 4 m². At the top of the stacks, as shown in Figure 12.4, are automatically controlled opening windows with a total free area of approximately 7.9 m². The opening area is roughly based on a rule of thumb for chimney tops that the area of the vertical opening faces should be at least twice the flue cross-sectional area. The discharge coefficient for windows varies from

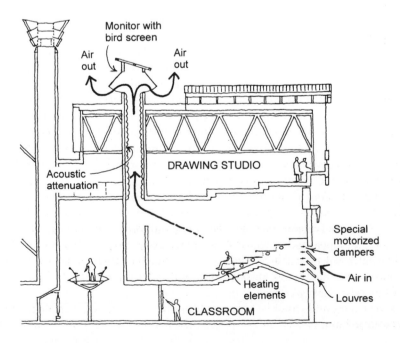

12.3 Air passage through auditorium.

12.4 Stacks at roof level.

approximately 0.57 at an opening of 60° to 0.63 at 90°. There is a slight overhang at the top of the stacks to help reduce the chance of rain entering.

The key points in the design of the auditoria are:

1. The spaces have high thermal mass and high ceilings. In summer the building structure is precooled at night by allowing air movement through the room. This helps to lower the day-time temperatures.
2. The air path from the air intakes, through the space and out through the stacks, is of low resistance, i.e. the pathway is mainly unobstructed.
3. The stacks terminate approximately 3 m above the roof line to avoid local turbulence. The air exhausts via automatic opening windows at the stack tops.

12.3 Lighting

Daylighting

The building has an extended perimeter to produce good daylighting for all spaces. In the early design days the architects tested a 1:50 scale model of the building under the artificial sky in the School of the Built Environment at De Montfort University and, to a lesser extent, tested a smaller scale model on a heliodon. In addition, daylight factors for some areas were determined using the Radiance daylight simulation model, again at the School of the Built Environment.

These studies resulted in adjustments to the window sizes and positions in some of the spaces and, in particular, a lightshelf was incorporated on the courtyard side of the electrical laboratories to reduce the possibility of glare on the computer screens from the high-level windows and to increase the daylighting levels in the centre of the rooms. The lightshelf was again modelled on the Radiance programme and increased daylight factors were predicted for the centre of the space compared with the initial model of the space without the shelf.

One of the courtyard lightshelves faces north, and on first appearance may seem to be unnecessary. However, the white panels which form the external walls of the electrical laboratory wings in the courtyard are quite reflective (section 5.4) and there is a significant quantity of light reflected from the south-facing courtyard elevation into the north side of the laboratory opposite.

On the south side of the central building the perimeter folds back on itself to form a series of courtyards, thus improving lighting levels in the classrooms and staff offices. The external courtyards may also be used as open-air classrooms in summer months. It could be argued that the extended perimeter required to maximize daylight acts against more traditional ideas of energy conservation by providing a greater external wall area through which heat can be lost in winter. However, the high standard of insulation provided to lower the U-value of the construction in some ways reduces the significance of this issue.

A conflict can also exist between the quantity of glazing required for optimum daylighting and the requirement to reduce solar gain by shading. In addition, if internal blinds or external shades are used to reduce glare and/or heat gain, ventilation by opening windows may be hindered. All these aspects must be considered at an early stage in order to reach the optimum solution.

Figure 12.5 shows light levels for the building.

Although there are light level sensors inside and outside the building, those inside will be influenced by any artificial lighting in the vicinity. This is a problem when trying to use internal light level sensors to switch off artificial lighting to save energy when daylight levels are high enough. Thus, only outside sensors are used to switch off artificial lighting in certain spaces. The level at which the artificial lighting is switched off is determined experimentally and iteratively. After the level has been set, the room users then provide feedback; if necessary, the level is altered and the process starts again. This exercise has yet to be undertaken fully within the building. However, initial experiments are being performed. The internal light levels shown in Figure 12.5 were taken while ensuring that the artificial lighting was off.

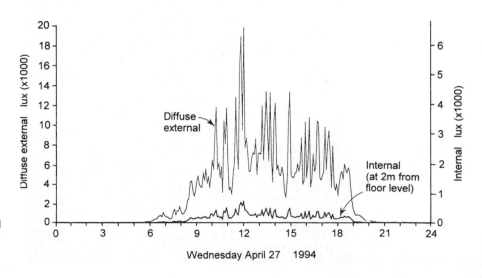

12.5 Measured external diffuse and estimated central concourse internal light levels.

Artificial lighting

Artificial lighting is generally fluorescent, using either strip lighting or low-energy compact lamps in the classrooms, auditoria, offices, drawing studios, electrical laboratories and circulation spaces. In the offices, teaching rooms and computer laboratories the overhead lighting levels were designed to give 300 lux on the working plane with sufficient sockets available for additional task lighting if required. High-frequency ballasts were incorporated in the fluorescent strip lights to reduce the flicker associated with standard lights of this type as the flicker can affect people with mild epilepsy, particularly when using computer screens. High-frequency ballasts also reduce the energy consumed by the lamp. However, a disadvantage is that high-frequency ballasts are expensive compared to standard ballasts, and this will no doubt remain until they become the industry standard.

In general, the artificial lighting is switched in rows so that those lights nearest the windows can be turned off independently when daylight at the perimeter is adequate. Passive infrared movement detectors can disable the artificial lighting when spaces are unoccupied.

The high spaces of the concourse (Figure 12.6), machine hall and central laboratory have high-bay high-pressure mercury lighting. These lamps were chosen for the machine hall and laboratory to provide the high lighting level of 1000 lux required

12.6 Concourse space through central building showing daylighting effect.

and also for their efficacy. There is no passive infrared control of the lighting in these two areas as they contain rotating machinery and it could be dangerous to machine operators if the light levels were to alter suddenly.

12.4 Acoustics

The following acoustic issues were addressed during the design phase:

Background noise levels in the auditoria

The air intakes to these spaces are from a busy street. It was agreed with the client that the required environment should provide a balance between the very low noise levels associated with air-conditioned spaces and the noise associated with opening windows in a more traditional classroom environment. The constraints on the solution are cost and space. To remove noise without introducing excessive resistance to air flow requires more space for either acoustically lined plena or acoustic splitters.

Both options were used in the auditoria. The lining is 50 mm Rockwool with a fabric covering to retain the fibres; the splitters are two banks of 500 mm deep, 50 mm wide and 2 m high perforated steel panels filled with Rockwool and suspended in the air intake path with 50 mm gaps in between. In addition, the stacks were lined with 25 mm of fabric-covered Rockwool to reduce potential disturbance from aircraft noise. The L_{A10} noise level (Appendix C) in the street outside the auditoria is 70–75 dBA. Initial sound level measurements have shown a reduction in these levels by 20–25 dBA, giving a background level of approximately 50 dBA just after the attenuators.

Reverberation times in the teaching spaces

The exposed surfaces of brickwork and concrete have low coefficients of absorption and there is therefore a tendency towards high reverberation times. The auditoria reverberation time calculations carried out at the design stage suggested that for a reverberation time of 1.2 to 1.4 seconds at 500 Hz significant amounts of acoustic panelling would be required. This was incorporated on the walls with the soffit left free to expose the concrete, thus taking advantage of its thermal mass. The panels are made of perforated medium density fibre board (MDF) behind which are 50 mm deep Rockwool cavity batts. There is a conflict between the ideal construction for lower peak temperatures and good acoustics, because the exposed surfaces of high mass – such as the brick walls and concrete soffit, which are desirable for their high admittance and can be used as a thermal sink for heat in summer – tend to produce high reverberation times as they have low sound absorption coefficients. A compromise has to be reached that will be acceptable for both criteria.

Noise transfer from inside to outside

The mechanical laboratories shown in Figure 12.7 are sited quite close to a row of domestic houses. Activities such as metal cutting and engine testing can generate internal noise levels in excess of 90 dBA. As a result, the mechanical laboratory ridge

12.7 Mechanical laboratory showing buttress side wall vents with classroom exhaust stack and boiler flue behind.

vents are open on two sides only (i.e. those facing away from the houses) and the throats of these exhausts are lined internally with a mineral wool quilt in a perforated metal casing. As an additional precaution the high-level gable glazing that faces the houses is triple glazed and fixed.

For this project there was only the standard requirement for acoustic separation between adjacent internal spaces. However, this could be another important aspect to be considered in certain situations. At De Montfort each space has its own separate ventilation system with separate exhaust paths, thus there is no risk of noise being carried from place to place by a common stack as often occurs with air-conditioning ductwork serving a number of rooms.

12.5 Controls

One of the main differences between De Montfort's engineering school and a naturally ventilated Victorian building is in the control. From the Victorian era manual controls have been used but often haphazardly. The school, on the other hand, uses a building management system (BMS). Room temperatures and the amount of air entering or leaving the spaces can be controlled using criteria such as:

- fresh air requirement of the occupants (carbon dioxide sensors)
- temperature (individual or averaging temperature sensors in spaces)
- rain (sensor on top of one of the stacks – printed circuit board)
- wind (rotating cup anemometer on top of one of the stacks)
- time (time clock control as part of BMS)
- air movement (flow meters in stacks)
- noise (sound level meter – not yet installed – future provision only)
- fire condition (as sensed by the fire alarm system throughout the building).

It is intended that the building be used as a teaching aid for students and also to further research into naturally ventilated buildings. To this end additional sensors have been incorporated such as relative humidity and temperature sensors within the layers of construction of an external wall, floor and roof. In addition, there are sensors measuring daylight level externally and lighting levels internally. These can be used for practical as well as experimental purposes by programming the computer to switch off the artificial lighting when external daylight reaches a predetermined level.

All control set points are adjustable via the computer but also in each space users can override the BMS using local controls.

The fire alarm overrides all the environmental sensors, shutting air intakes and opening the exhausts. There is also a manual switch for fire-fighting personnel which allows exhausts to be left as dictated by the BMS or closed as required. Sometimes fire officers take the view that exhausts should be opened to allow smoke to escape. At other times they may wish to close the exhausts to starve the fire of oxygen.

In terms of the day-to-day running of the building, there are four basic conditions for which the controls must allow. The auditoria are the most complicated areas and, using them as the example:

1. In summer, if the auditorium is occupied and the inside temperature is higher than the outside temperature (and higher than the internal set temperature) the air inlet and exhaust dampers will be fully open. However, if the internal temperature is lower than the external temperature the dampers will open to a lesser degree based on the reading from the carbon dioxide sensor. Additionally, if there is rain accompanied by a high wind speed, exhaust dampers will close to a minimal open setting. In extreme winds the air intake and extract dampers will close completely, overriding the carbon dioxide control. The fan will be brought into operation to lower the internal temperature when required.
2. At night, in summer, the external air flows through the space to precool the structure to a preset temperature above the dewpoint (temperature and relative humidity sensors monitored) and no lower than the temperature that would necessitate reheating in the morning for the first lecture. This is set at 17 °C at present. Air flow will be assisted by the fan if necessary.
3. In winter, when the room is occupied, the vents open under the control of the carbon dioxide sensor. As for summer conditions, combinations of wind and rain close the air exhausts and intakes to differing degrees related to severity.
4. At night in winter, or when the rooms are unoccupied, air intakes and exhausts are closed to avoid unnecessary heat loss.

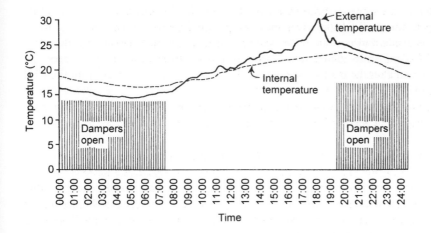

12.8 Recorded temperature data during a heat load test for auditorium 1 on 27–28 June 1994.

Figure 12.8 shows the outputs from the sensors during a heat load test in one of the auditoria for a 24-hour period where the external conditions reached 30 °C and the internal heat gains were 16 kW. It can be seen that the internal temperature did not rise above 25 °C, which is significantly better than predicted by the early models and calculations (Table 12.4). One likely reason for this is the effect of the wind. The auditoria are currently the subject of extensive monitoring by De Montfort. Results will be published in due course.

12.6 Other aspects

CHP unit

The analysis of the building loads at the design stage showed that large-scale combined heat and power (CHP) plant would not be appropriate as there is little use for large quantities of heat in the summer. A small-scale gas-fired unit was incorporated which produces approximately 40 kW of power and 80 kW of heat. The heat meets the domestic hot water load in the summer months. In winter the CHP unit acts as the lead boiler with a condensing boiler (300 kW duty) as second on line. Two gas-fired high-efficiency boilers (each rated at 300 kW) make up the remaining full heating load. The controls are set to operate to minimize running costs. Thus, if there is no requirement for heat but a requirement for power, the unit will not run and, instead, power is supplied more economically from the grid. Similarly, if there is no power requirement, but a heating requirement the unit will not run and, instead, heat is supplied more economically from the boilers.

The CHP unit does not at present act as a standby generator, although at some time in the future it may be adapted to serve this purpose. The unit has a water-cooled reciprocating engine based on models used in tractors. Heat is reclaimed from the engine jacket, the lubricating oil and the exhaust flue gases to warm the hot water which is fed into the hot water heating system via a water-to-water heat exchanger within the CHP unit. The power generated is run onto the main power intake for the building and is synchronized with the mains electricity supply to ensure that all the power is in phase. If the power of the unit is not being fully used it returns to the mains and feeds the local electricity supply grid. As a safety precaution, therefore, the CHP plant shuts down if the main supply shuts down. This ensures that anyone working on the mains will not find them live through backfeed from the CHP plant.

A further reason for incorporating the CHP unit was to make it available for study by the students of the engineering school.

Solar panels

Some thought was given at an early stage to the incorporation of solar panels to preheat the hot water service. However, the decision to provide a small-scale CHP unit meant that the solar panels would not be a viable option as both items of plant would have been fulfilling a similar function.

Heat recovery

Heat has not been reclaimed from the stack exhausts. Theoretically, it is possible, but the additional resistance imposed by a run-around coil or air-to-air heat exchanger would be likely to impose a requirement for a larger fan in regular operation, and the viability of this would be questionable. In addition, the geometry of the building, and, in particular, the relationship between the air intakes and the exhausts in the auditoria, were not ideally suitable for reclaiming heat from the exhaust and putting it back in at the intake to preheat cold incoming air in winter months. However, heat recovery can be desirable and should be incorporated where appropriate in other designs.

Costs

One of the main questions that tends to be asked about this building is whether it cost more, or less, to build than a more standard engineering school with full or part air-conditioning. In part response to this question, Table 12.5 shows some of the figures produced by the quantity surveyors.

The mechanical and electrical services capital costs are obviously much lower, and we also expect running costs to be much lower. Some of the cost of plant has been shifted towards the building.

To reduce costs a specific attempt was made to make aspects of the building relating to natural ventilation serve more than one function. For example, the auditoria stacks provide the main exhausts for air from the auditoria but, in addition, as mentioned above, the structure of the stacks supports the drawing studio roofs above and the glazed tops allow some daylight into the internal parts of the auditoria.

Table 12.5 Approximate capital cost comparison between two comfort cooled auditoria and two naturally ventilated auditoria at De Montfort

	Comfort cooled (£)	Naturally ventilated (£)
1. Comfort cooling plant	70 000	–
2. Plant room space	30 000	–
3. External intake louvres	–	9 000
4. Attenuators and controls	–	11 000
5. Plenum duct space	–	5 000
6. Ventilation stacks (4 No.)	–	35 000
7. Attenuators, ventilation terminals and controls to stacks	–	32 000
	100 000	92 000

Project principals

Client	De Montfort University, Leicester
Architects	Short Ford & Associates
Services engineers	Max Fordham LLP
Structural engineers	YRM/Anthony Hunt Associates
Quantity surveyors	Dearle & Henderson, London
Landscape architects	Livingston Eyre Associates
Main contractors	Laing Midlands
Mehanical contractors	How Engineering
Electrical contractors	Hall & Stinson

References

1. Data provided by B. Ford (Short Ford & Associates).
2. Lane-Serff, G., Linden, P., Parker, D.J. and Smeed, D.A. (1990) *Laboratory Modelling of the New Engineering School at Leicester Polytechnic*, University of Cambridge Department of Applied Mathematics and Theoretical Physics in conjunction with Cambridge Architectural Research Ltd.
3. Letter of 23 August 1990 from Dr K. Lomas, De Montfort University School of the Built Environment, to A. Short.

Further reading

Anon. (1997) *The Queens Building*, De Montfort University, Department of the Environment. New Practice Final Report 102. BRE, Garston.

Bunn, R. (1993) Learning curve. *Building Services*, 15(10), 20–5.

Hawkes, D. (1994) User control in a passive building. *Architects' Journal*, **10**(199), 27–9.

Stevens, B. (1994) A testing time for natural ventilation. *Building Services*, **16**(11), 51–2.

Swenarton, M. (1993) Low energy gothic: Alan Short and Brian Ford at Leicester. *Architecture Today*, **41**, 20–30.

CHAPTER 13

The Charles Cryer Studio Theatre

This arts centre in Carshalton, London (shown in Figure 13.1), started with a brief for a low maintenance, environmentally friendly design and a specific requirement to provide comfort cooling to the theatre. To achieve an acoustic performance of NR25 (Appendix C) within the theatre space, external noise from the adjacent busy main road had to be excluded and the proximity of local housing meant that performance sound had to be kept in at night. The combination of an acoustically airtight theatre,

13.1 Front elevation.

250 people and stage lighting heat gains, made ventilation and cooling essential. A range of options was considered including natural and mechanical ventilation, remote-sited air condenser and various water-based cooling alternatives. The restricted space in the listed building envelope did not allow suitable plantroom space either inside or outside the building for a traditional air-condenser system. Acoustic enclosures around such a system would have required a total volume of 20–30 m.[3]

The solution finally proposed and constructed was to use the rather novel solution of groundwater as a cooling medium. The advantages and disadvantages of groundwater are shown in Table 13.1.

There are many potential sources of groundwater, and many methods of extraction and ways in which the water can be used. The first step was to survey the surrounding area, study geological maps and discuss the options with the National Rivers Authority (NRA),[1] local water authority and the British Geological Survey (BGS).[2]

The NRA and BGS, with their extensive knowledge of a given area, were quickly able to determine the probability of suitable groundwater sources.

This research will generally provide an overview of the local geology, areas of surface water, underground water, location and extraction capacity of the local boreholes and the seasonal and secular (i.e. yearly) reliability (variation) in water levels.

For the Arts Centre the most suitable and reliable method of extraction was considered to be a deep borehole to tap the water from an underground aquifer. Substantial amounts of surface water appeared to be available locally but the source was unreliable.

The NRA and BGS data suggested that an abstraction borehole drilled 50 m deep through the chalk strata and into the underground aquifer would provide a reliable water table well (Figure 13.2).

Similar boreholes in the area suggested that the natural water level in such a borehole would be within 5 m of the ground level with seasonal variations of up to

Table 13.1 Groundwater as a cooling medium

Advantages
- Plentiful quantities of cooling medium
- Minimal plantroom space
- Low maintenance
- Low running cost
- No CFCs or HCFCs
- Simple and traditional equipment and technology
- Recyclable resource

Disadvantages
- Relatively high coil-on water temperature (i.e. the temperature of the water when it enters the cooling coil of an air-handling unit) compared with a conventional air condenser cooling system (i.e. one where the heat is rejected at an air-cooled condenser)
- Information regarding available groundwater and yield may be limited for a particular site
- Generally more expensive than a conventional air condenser cooling system unless 25 years' life cycle costing is considered.

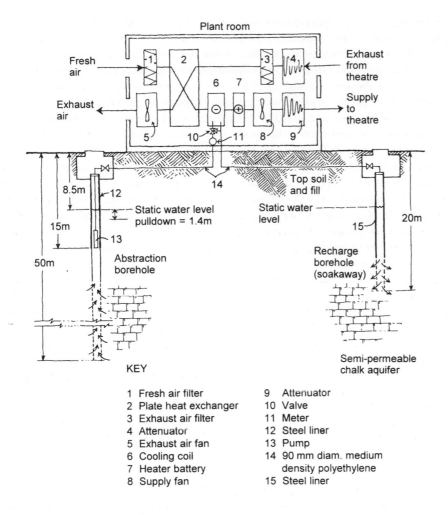

13.2 Borehole cooling and ventilation schematic.

KEY

1 Fresh air filter
2 Plate heat exchanger
3 Exhaust air filter
4 Attenuator
5 Exhaust air fan
6 Cooling coil
7 Heater battery
8 Supply fan

9 Attenuator
10 Valve
11 Meter
12 Steel liner
13 Pump
14 90 mm diam. medium density polyethylene
15 Steel liner

1 m. Yield was expected to be more than adequate and a water temperature of 10 °C was normal for similar 50 m deep boreholes in the area – at this depth the temperature appears to be constant. The strata of the chalk aquifer generally provides clean, clear potable water. Although the water is very hard (i.e. has a high content of calcium and magnesium bicarbonates, in this case 220 mg/1, when expressed as $CaCO_3$), the water in the heat exchanger would be less than 20 °C, which should not cause fouling of the waterways.

To conserve and recycle this natural resource it is necessary to ensure that the water is not contaminated in use and that it is returned to the aquifer. This is done, in part, by separating the supply and return water so that the return water does not raise the temperature of the supply water. Normally, a separation of 100 m is recommended, but this was not feasible in this case and so a second 20 m deep borehole (soakaway), which returns water to the aquifer, was located 40 m distant from the first. To provide additional security the boreholes are set at different levels in the aquifer and with the abstraction borehole on the upstream side of the nominal direction of flow of the aquifer.

A multi-stage borehole pump installed inside the borehole (and so out of sight and requiring no plantroom space) raises water from the borehole to the airhandling unit, which is 10 m above ground level.

Although the current water resource is considered to be plentiful and recycled, the water extracted is metered and the annual quantities 'borrowed' are restricted by the NRA. In order to minimize the pumped water demand and therefore energy used and running costs, suitable measures have been taken to reduce the cooling demand.

The building relies on the heavy thermal mass of the fabric, precooling and free cooling (Chapter 12) to reduce the cooling loads. Free cooling uses the external supply air to help to cool the building (without running the borehole water pump) when the external air temperature is lower than the space design temperature. Two speed ventilation fans ensure that fan power can be minimized when the full cooling capacity is not required. When the full cooling load is expected, the management can use the manual override controls to precool the theatre to 18 °C and rely on the thermal mass of the exposed heavy building fabric to maintain comfort conditions for a limited period.

The mechanical ventilation system is a full fresh air (no recirculated air or bypass dampers), two-speed, ducted supply air and exhaust system with a heater battery, cooling battery, silencers, filters and plate heat exchanger for heat recovery. Fresh air enters the system from the north side of the building where the tree-lined courtyard air is cooler and less polluted and the ambient noise is lower. Fire dampers, bypass and air flow regulation dampers were designed out of the ventilation system to reduce maintenance. The measures taken to achieve low maintenance and reliability do, however, compromise the ideal conditions required for low cooling loads. The permanently in-line plate heat exchanger reduces the potential for free cooling, but in practice this is not a problem. Decorative finishes inside the building reduce the thermal mass available by insulating the heavy structure and building fabric from the space. Nevertheless, the measures taken reduced the design cooling load to 395 W/m^2. The system performance data is given in Table 13.2.

Table 13.2 Theatre ventilation system performance data

Full speed fan volume	4.2 m³/s
Fan power	7.5 kW
Air on-coil	26 °C dry bulb; 19 °C wet bulb
Air off-coil	13.8 °C dry bulb; 13.4 °C wet bulb
Cooling water on-coil	11 °C
Cooling water off-coil	18 °C
Cooling water flow rate	2.8 l/s
Cooling coil rows	8
Cooling duty	83 kW
Maximum capacity (persons)	250
Ventilation rate per person (maximum)	17 l/s
Air changes per hour (maximum)	12

With the peak cooling water flow rate established from the cooling duty as 2.8 l/s, the borehole diameters could be sized to achieve the necessary abstraction and recharge duties. The NRA and BGS can provide the necessary background information for this assessment. Table 13.3 lists the relevant borehole statistics.

The abstraction borehole has a mild steel lining tube (10 mm thick) for the top 15 m. The inlet to the pump is 15 m below ground level with a steel discharge pipe running to a flanged base plate in the abstraction manhole.

All pipework from the plantroom to each borehole is laid to fall to the boreholes and provided with a vent at the highest point in the plantroom. The discharge water from the cooling battery returns to the recharge borehole by gravity. When the borehole pump is not running, the water in the supply pipework drains down into the borehole so preventing any danger from frozen pipes, leaking joints or stagnant water.

Fully automatic or manual control is available to the theatre management. Under automatic or manual cooling mode, averaging thermostats will operate the borehole pump when the room air temperature exceeds a predetermined level. A motorized three-port valve in the borehole supply pipework then controls the volume of borehole water available to the cooling battery. A further thermostat in the supply air ductwork controls the motorized three-port valve to ensure that the minimum supply air temperature does not fall below 14 °C.

Each borehole cost approximately £6000, but the complete package – including boreholes, pipework, pump, controls and manholes – is nearer £30 000. With 83 kW of cooling this is over £360 per kilowatt. However, the installed system could easily provide 160 kW of cooling for little more than the cost of a higher duty pump, thus bringing costs down to about £200/kW.

If a traditional air condenser package had been feasible, the installed capital cost would have been approximately £15 000. Such a comparison is not strictly accurate, of course, as there are many other costs and considerations that should be taken into account, such as maintenance costs, running costs and the effect that the relatively high cooling medium water temperature has on the sizing of the air-handling unit.

Table 13.3 Borehole statistics

	Abstraction borehole	Recharge borehole
Nominal diameter (mm)	300	375
Depth (m)	50	20
Pumped flow (m³/h)	10	10
Pulldown[a] (m)	1.4	N/A
Steel liner depth (m)	15	15
Static water level below ground (m)	8.5	8.4
Pump inlet level below ground (m)	15	N/A
Variation in static water level (m)	1	1

[a] Pulldown is the difference between the water level in the borehole when the pump is running and when it is not.

Underground aquifers can provide water at temperatures as low as 10 °C but by the time this reaches the cooling coil of an air-handling unit the water on-coil temperature is likely to be about 12 °C. It is then difficult to achieve air temperatures off the coil lower than about 14 °C. As a result, the air-flow rates, air-handling unit size, fan size and ductwork may need to be increased to achieve the required cooling duty.

Maintenance costs on the borehole have proved to be extremely low. The pump and the motorized valve are the only moving parts. Borehole water can deposit a slime within the pipework and cooling coil. To allow any such residue to be cleaned out the cooling coil is provided with a flushing facility.

Running costs are low as only the 2.5 kW pump, 65 mm motorized valve and the controls require power. The coefficient of performance (COP) of the system is very good. With the fans running at full speed the electrical power absorbed by the fans, pump and controls is approximately 12 kW, and this provides a cooling output of 83 kW.

A number of variations and modifications to this scheme were considered (e.g. run-around coils, recirculation dampers, evaporative cooling, water-to-water chiller) to improve performance, efficiency and/or energy consumption but all these variations would have had an adverse effect on either capital cost, maintenance, simplicity or reliability.

Summary

Groundwater is an environmentally friendly answer to many cooling needs and this system at the Charles Cryer Studio Theatre has proved to be very simple, reliable and economical to operate and maintain. If groundwater is available locally, and if demand is not oversubscribed, such cooling may well be a feasible option.

Project principals

Client	London Borough of Sutton
Architects	Edward Cullinan Architects
Services engineers	Max Fordham LLP
Structural engineers	Jampel Davison & Bell
Quantity surveyors	Dearle & Henderson
Main contractors	Eve Construction
Mechanical contractors	JW. Stubberfield & Sons Ltd
Electrical contractors	RTT Engineering Services

References

1. NRA is now the Environment Agency, www.environment-agency.gov.uk
2. British Geological Survey, Maclean Building, Crowmarsh Gifford, Wallingford, Oxfordshire OX10 5BB.

Further reading

Todd, D.K. (1959) *Ground Water Hydrology*, Wiley, Tokyo.
Wright, P. (1992) Dramatic effect. *Architects' Journal*, **16**(195), 36–45.

CHAPTER 14

The Environmental Building, Building Research Establishment

14.1 Introduction

Sited 300 m from a major motorway (the M1) in Garston, about 15 km north-west of London, is one of the most innovative buildings in the UK.

Its brief, which resulted from collaboration between the Building Research Establishment Energy Conservation Support Unit (BRECSU) and the Energy Efficient Office of the Future (EOF) Group, called for a landmark building combining the highest architectural standards with the latest innovations in energy efficient design. The building also needed to be environmentally friendly (a BREEAM – Building Research Establishment Environmental Assessment Method – rating of excellent was required), to act as a test bed for BRE research, and to demonstrate techniques which could be adopted in the design of future offices.

Spatially, it had to provide 1300 m^2 of offices for about 100 staff and 800 m^2 or so of seminar and associated facilities. The offices are 30 m × 13.5 m on plan with the short axis running almost due north–south. The grid consists of a 7.5 m southerly zone, a

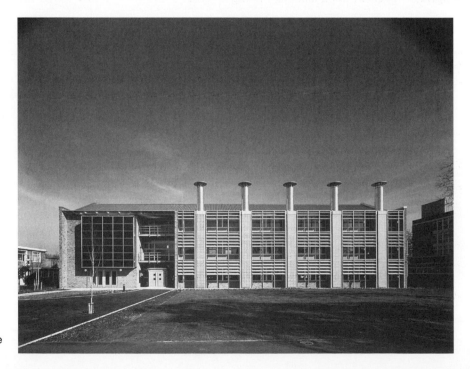

14.1 South façade of the building.

1.5 m corridor and a 4.5 m northerly zone. The ceilings of the ground and first floors are 3.4 m high at their peak. On the uppermost second floor the sloping roof varies from 2.5 m to 5.0 m in height.

The seminar facilities consist of a main space to seat 100 people and two smaller rooms for about 20 each.

To the south of the building there is an open space of about 30 m; thus there was ample opportunity to use solar energy. Ventilation strategies in the built environment generally depend on noise levels (discussed on p. 204) and site wind patterns, to name two of the more important factors. The site is fairly open and consists mainly of two- or three-storey buildings with the occasional block of five or six storeys. Although winds come from all directions, winds from the south-west are somewhat more common than others. Figure 14.2 shows the wind distribution for a representative year for January to December.

Geologically, the site consists of made ground over glacial sand and gravel. At a depth of about 22 m, the Upper Chalk is reached and after about 30 m more, the Middle Chalk. The chalk is part of a large aquifer in a broad area in southern and eastern England, including London, with an important potential for borehole cooling systems as used at the Environmental Building (EB) and discussed below.

Perhaps the greatest significance of the EB is that it is one of the first buildings to result from a holistic view of construction. It takes advantage of all the available local sources of energy (sun, wind and groundwater), and in addition goes beyond this by incorporating a philosophy of which materials should be used and how the building should be constructed (and deconstructed). In the next 20 years every aspect of the EB will be improved upon, but it will have more than fulfilled its purpose in setting an agenda for architectural and engineering quality.

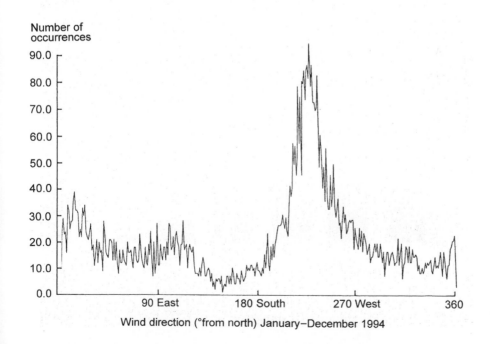

14.2 Wind distribution by direction for January–December 1994 at the BRE.

Wind direction (°from north) January–December 1994

14.2 Energy use, comfort and form

Energy consumption was targeted at 47 kWh/m^2 yr for delivered gas and 36 kWh/m^2 yr for delivered electricity; this criterion was based on a 30% improvement in best practice in 1994/95 when the EOF brief was being developed. The corresponding total CO_2 figure was 34 kg/m^2 yr. Since electricity consumption produces about three times at much CO_2 as gas use (see Table 7.4), developing the design is partly a matter of playing off one factor against another. For example, more glazing results in more natural light; thus, in principle, less electricity is needed for artificial lighting and less CO_2 is produced. This is balanced against more CO_2 resulting from burning gas for increased heating (due to glazing being worse for insulation than typical wall constructions).

To put the above figures in perspective, Figure 14.3 shows the very encouraging downward trend in CO_2 production (and corresponding energy use) that has been achieved over, say, the past twenty years.

One of the most stimulating aspects of the brief for the design team was that in addition to limiting energy use, comfort criteria were set of not exceeding 28 °C in the offices for more than 1% of yearly working hours, i.e. 20 hours and not exceeding 25 °C for more than 5% of yearly working hours (the 'year' in question is discussed below). The importance of these two criteria is that they provide a clear-cut approach for designers – any design proposal can be tested to ensure that both conditions of the brief are met. It has been said that the comfort temperatures are somewhat arbitrary and numerous alternatives could be proposed, but the key point is to have such conditions.

The design team then considered what form of building construction could meet the brief. It was agreed that solar energy and daylight should be used as much as possible to reduce the energy demand in winter and the need for artificial lighting

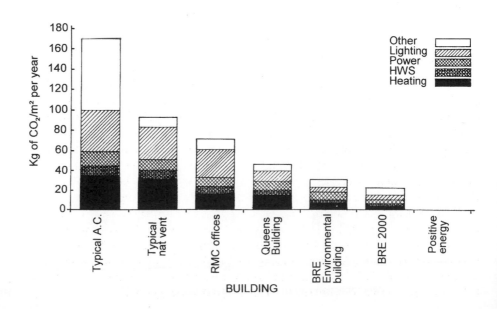

14.3 Annual CO_2 production for various buildings and types.

throughout the year. This, along with site constraints, led to a narrow plan, three-storey building with generous floor to ceiling heights and a long south-facing façade with about 50% glazing. More glazing on the south side would have tended to increase the hours above the target temperatures.

It was clear that it would be difficult to meet the energy criteria if the building were sealed and air-conditioning were used. Thus, it became critical to ensure that the offices had sufficient thermal mass and a ventilation strategy to deal with heat gains during the day and allow for cooling at night using lower temperature air. It was decided to use an exposed concrete slab (discussed in more detail below) on the ground and first floors and timber boarding on the top floor. Timber was selected for a number of reasons, including aesthetic appeal, the use of a renewable resource, and the avoidance, on both structural and cost grounds, of heavy concrete in the roof.

External walls are made of 100 mm brickwork, 100 mm of blown Rockwool insulation and 150 mm blockwork with a dense plaster coating giving a U-value of 0.32 $W/m^2 K$. The floor is a partially insulated concrete slab with a U-value of approximately 0.33. For the roof, with its 75 mm of timber, 200 mm of Rockwool and aluminium cladding, the U-value is 0.16. Finally, argon-filled, double-glazed windows with a low-emissivity coating have a U-value of 2.0.

These elements of construction and form (and numerous other alternatives) were simulated using computer programs to test compliance with the brief. Weather data from London for a 20-year period was analysed and the years ranked on the basis of the average temperature during the three summer months. The fifth year in the ranking, 1994, was selected according to a variation of the 'Goldilocks and the Three Bears' rule: not too hot, not too cold, but just right. This data for temperatures and solar radiation was used in connection with the given occupancy levels, an assumed lighting load of 10 W/m^2 and an assumed power load of 15 W/m^2. The results of a number of studies by EOF group members and us indicated that the design would meet the brief.

14.3 Ventilation strategy

Ventilation is a crucial element in the design of low-energy buildings. In the 1960s and 1970s one of the reasons given for preferring mechanical ventilation and air-conditioning was that natural ventilation could not be controlled and consequently comfortable conditions and energy conservation could not be provided. This argument persists, for example, in France, where assisted natural ventilation buildings (other than homes) are rare. What we have seen recently is the development of more tightly sealed buildings, defined ventilation routes and control techniques which allow the required air to be introduced as and when needed, day and night.

For the EB, there were several starting points. The first was to minimize the energy use associated with ventilation. This favoured either cross ventilation or stack-effect ventilation (as in the De Montfort Queens Building – see Chapter 12) or a combination of both; in addition there was the possibility of providing some low-powered mechanical ventilation in the form of, say, a simple fan (see Figure 9.21). It also favoured high thermal mass to smooth cooling (and heating) loads.

A second point was to link the ventilation path with the thermal mass to take advantage of night cooling, and a third was that the normal requirement that a space for heating pipes, electrical cables and so forth had to be found. A further key consideration was that a percentage of cellular offices had to be allowed for and so a way of maintaining air flow through open plan offices was required.

Close co-operation among the architect, structural engineer and environmental engineer led to an attractive, structurally efficient sinusoidal slab as shown in Figure 14.4. The floor is a composite construction incorporating a pre-cast concrete former and a profiled *in situ* topping. Insulation below the screed helps to create some

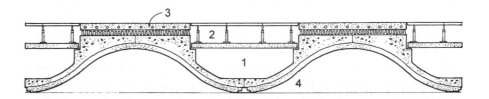

KEY 1 Ventilation paths through slab serving floor below
      2 Raised floor zone for cables and pipes for floor above
      3 Screed with heating/cooling pipes and insulation beneath
      4 Pre-cast concrete (75 mm) ceiling with *in situ* concrete topping
      5 High level windows
      6 Side-hung casement windows
      7 Bottom-hung translucent windows

14.4 Sinusoidal slab section and view of a typical bay.

thermal separation between the two floors. The floor is designed as a folded plate with, typically, the bottom of the 'wave' in tension and the top in compression. An incidental advantage of the sinusoidal slab was that its greater area provided a significant increase in thermal mass.

The ventilation paths were studied for a variety of situations including combinations of summer/winter, day/night, wind/no wind, wind from the south/wind from the north, rain/no rain and open plan/cellular offices. Figure 14.5 gives an example.

An essential element of the ventilation strategy is that it is 'loose-fit' and has numerous air paths. In the summer cross ventilation is facilitated by the shallow floor plan. Air can come in under BMS control through openings at the edge of the slab ('1' in Figure 14.4) or through the high level windows ('5' in Figure 14.4), or the side-hung windows can be opened manually. In areas with cellular offices, the sinusoidal slab lets air pass over them thus permitting ventilation of the open plan areas. The cellular offices themselves are principally ventilated from one side by their windows.

The vertical shafts on the south façade shown in Figure 14.1 allow stack effect ventilation. They operate simply as stacks in no wind conditions – when there is wind it will induce flow up the shafts, thus assisting the stack effect. Terminal design, therefore, needs to take account both of no or low wind conditions and more windy

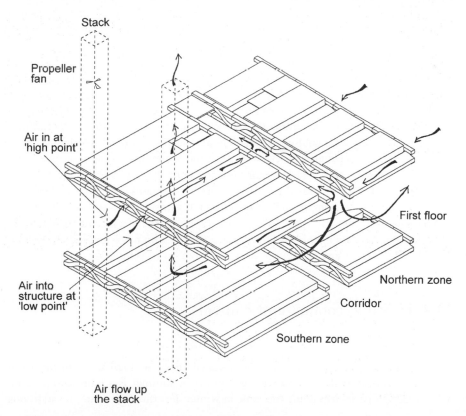

14.5 Ventilation path in open plan offices on a summer day with no wind.

ones. For a good discussion of the principles involved, with some testing results, consult reference 1.

Propeller fans of 48 W in the stacks are available to provide mechanical assistance, either in low wind conditions during the day or to assist night–time cooling if needed; the duty of each fan is about 0.33 m^3/s in 'free air' (i.e. against a negligible resistance). It is worth while noting that meteorological wind speed data which indicates few prolonged calm periods in the UK may mislead designers. The data often comes from exposed sites and is usually taken at a height of 10 m, whereas buildings tend to start at the ground and may be surrounded by other similar structures.

The main principles of the ventilation strategy were confirmed by studies carried out using a plexiglas model immersed (upside down) in a tank.[2] The technique is known as salt–water bath modelling because the flow of small quantities of salt water is used to simulate heat flows caused by gains from, for example, occupants and equipment.

During the winter, ventilation is, of course, not needed to reduce temperatures and the required air quantity is related to stuffiness and odour control. At the EB air is provided by the building management system (BMS) automatically opening the high level windows a small amount. This can be supplemented by the users manually opening windows slightly if desired.

The BMS controls the high level windows and the glazed openings into the slab, the translucent openings from the offices into the stacks and the stack fans. From the outset it was considered essential for the occupants to be able to control their environment and, in practice, this has taken the form of a number of hand–held infrared controllers which allow adjustment of ventilation openings (and solar shade louvre positions and lighting levels as discussed below).

Ventilation air flows thus simply depend on varying opening sizes and occasionally running fans. No filters are used. With regard to noise, during the day the levels outside vary from about 45 to 55 dBA, with the motorway being an important source. Generally, noise coming in via open windows has not been a problem.

Whether the glass block ventilation shafts are 'solar stacks' is the subject of mild debate. The effective functioning of the system does not require any solar assistance and, indeed, translucent glass blocks were only one of a number of options considered for construction including rather more opaque concrete and metal. The technical issues that need study include the increased direct solar gains through the blocks into the space that one is trying to cool and conduction gains and losses through the back wall of the stack into the space. The construction of the stacks and their number favours experimentation and it is hoped that research which is being carried out will determine if there is any significant solar contribution to ventilation rates.

14.4 Heating, cooling and controls

Heating

In the modern office building with a well-sealed envelope, high levels of insulation and significant internal gains, providing comfort in summer is a more complex and difficult design issue than doing the same in winter. For the EB, in fact, no additional heating is required during the day above an external temperature of about 7 °C. This is the same as saying that above 7 °C the heating is provided by the lights, occupants, equipment and solar gain.

Nonetheless, heating, of course, is essential at low temperatures and also for start-up situations. Gas was available on the site so the decision to use gas-fired boilers was fairly straightforward. One condensing boiler and one conventional high efficiency boiler are used, with the former sized at 40% of the load. This is a reasonable compromise between increased efficiency and the higher capital cost of a larger condensing boiler.

The choice of the main form of heating swung towards underfloor heating coils when the borehole on the site became likely (see below). This allowed the one system to perform the two functions of heating and cooling. Radiators could not have been used to provide the cooling as damaging condensation would have formed on their surface. The underfloor polybutylene coils are fed indirectly from the gas-fired boilers using an injection system which maintains a lower, weather-compensated, flow temperature to the coils. The underfloor heating system also has the advantage that, as it runs at a lower temperature than the rest of the system (i.e. the radiators and heater batteries), it produces a lower return water temperature which increases the efficiency of the condensing boiler.

There are radiators in almost all areas of the building, and there are heater batteries on the fresh-air inlets in all of the seminar rooms. In the offices the radiators have TRVs and there are also wall-mounted room temperature control/sensors which allow fine adjustment.

In the office areas only 38% of the floor area is taken up by the underfloor heating coils; the rest of the area is used as a distribution route for the piped and cabled services. This was to allow the client flexibility within the office spaces but, as a consequence, there was not sufficient area of underfloor heating coil to obtain the required output and so a supplementary form of heating was needed. This was provided in the form of quick-response perimeter radiators (mentioned above) affording further flexibility in control and comfort.

Figure 14.6 shows a simplified schematic of the heating and cooling system which concentrates on the offices area and omits the radiator circuits for clarity.

In sizing the heating system a ventilation rate of one air change per hour was allowed for. Every effort was made during the design and construction period to produce a well-sealed building which would be capable of fine control in the 'winter' (in the 'summer' this is much less critical because one is often looking for higher ventilation rates). The building was pressure tested at a pressure of 25 Pa and found to have an air leakage index of 8.2 m^3/h per m^2 of envelope area, putting it between the categories of 'tight' and average. Further testing is expected to be carried out to see if any important leakage paths can be identified.

The heating/ventilation system in the offices is a once-through system where the incoming fresh air is under 'loose' control by the BMS. An alternative winter ventilation source of manually operated trickle ventilators situated at high level to ensure mixing of the incoming cold air with room air was considered. However, it was thought that since the control system would cater for the ventilation needs and would have the further advantages of tempering half of the fresh air by leading it through the slab openings, trickle units were not appropriate in this instance.

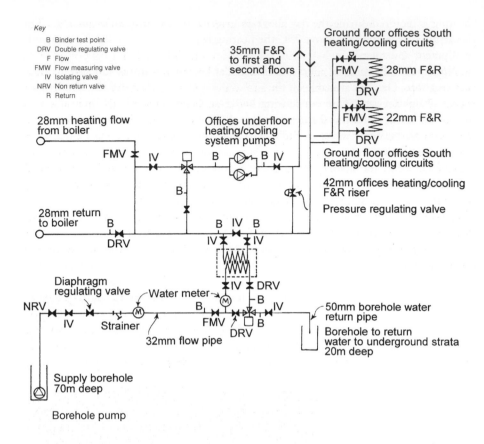

14.6 Simplified heating and cooling schematic.

Mechanical ventilation with heat recovery was also considered. We have seen that the energy consumption of the building is quite low and that in fact no additional heating during the day is required above an external temperature of 7 °C. Thus, the use for any heat which might be recovered is limited. Additionally, the cost of a mechanical ventilation system with heat recovery which is only used a small percentage of the time (namely, during the occupied daytime period when the outside temperature is below 7 °C) has a very long payback period – here, it was over 100 years. In other situations, though, where, for example, noise is an issue, mechanical ventilation might be necessary; in such cases, heat recovery is likely to be economical.

Cooling

Computer simulations indicated that it would be possible to meet the requirements of the brief for the offices area without any additional cooling (and this indeed proved to be the case – see Section 14.9), but for the main seminar room it would be much more difficult. A number of different options including a cooling coil linked to a dry air cooler were studied and a borehole system was selected as most in keeping with the overall project philosophy. It then made sense to link the borehole system to the offices' area pipework to provide a low cost improved environment that bettered the brief. Computer and salt-water bath studies both indicated that it should be possible to achieve a 2 °C lowering in peak temperature and this was judged sufficiently encouraging to proceed.

Figure 14.7 shows a borehole drilling rig on the site. The 150 mm supply borehole into the Middle Chalk is 70 m deep and provides 1.5 l/s of water at about 11 °C. It is pumped through a stainless steel heat exchanger and then discharged at 16 °C to a second 20 m deep borehole 60 m from the first. Permission for the borehole was negotiated with the National Rivers Authority, one of whose prime concerns was to ensure that the only change to the water was the necessary temperature increase. Consequently, it was not acceptable to them to use the water in a fountain or other water feature for fear of surface contamination. The discharge borehole only needed to be 20 m deep on the grounds that any possible pollution would be filtered out by the chalk before reaching the rest water level.

Considerable attention was given to control and comfort. To reduce the risk of discomfort which might result from the floor above wanting cooling with the floor below not needing it, insulation was incorporated below the screed containing the pipes (see Figure 14.4).

14.7 Borehole drilling rig.

Calculations were also carried out to estimate the temperature at carpet level to guard against complaints of 'cold feet'. On a hot summer day with an internal temperature of, say, 25 °C, it was calculated that the carpet temperature might be 21 °C or so. This should be acceptable but, as further insurance, manual on–off valves were provided to isolate flow to a bay in the event that individuals found the floor too cool.

The cooling effect of the floor is principally (about 70%) due to radiation but there is also a convective effect. The total is about 7 W/m^2 per degree of temperature difference.

The ventilation system is unchanged whether there is heating, cooling, or neither, and so remains simple (if somewhat less efficient) – there is no recovery of 'cool' thermal energy. This combination of ventilation and cooling is low cost and CFC and HCFC free.

Controls

The BMS controls the ventilation, heating and cooling. The backbone of the system is a network which allows for the integration of different services – for example, at the EB, the BMS is by one manufacturer and the lighting control system by another.

The control algorithms allow for four basic situations:

1. Winter day time: Minimum fresh air, heating as required.
2. Winter night time: Seal the building and maintain the temperature at the setback position.
3. Summer day time: If the external temperature is lower than the internal, freely ventilate the building and bring on the cooling if required. If the external temperature is higher, use minimal ventilation and, as before, cooling if required.
4. Summer night: Ventilate the building (i.e. 'night cooling') as needed to reduce the temperature (providing, of course, that it is cooler outside). In order to avoid the risk of having to bring the heating on the next morning the temperature is not normally dropped below 17 or 18 °C.

Extensive data logging capabilities have also been incorporated in the system.

14.5 Control of solar gain and daylighting

Control of solar gain, daylighting, ventilation and thermal mass all go together. In many ways the starting point for low energy buildings is to make the best use of solar energy and daylight while ensuring that potentially deleterious consequences are avoided. In the summer the danger is potential overheating, dealt with by shading, ventilation and thermal mass; in the winter it is excessive heat loss, which can be managed by glazing systems with low U-values or by insulating 'shutters' at night.

At the EB, only the south façade needed extensive control of solar gain. This was achieved firstly by limiting the glazed area as noted above to about 50% after extensive computer simulations of heat losses, heat gains, daylight factors and lighting gains plus consideration of possible overheating. With greater solar control to maintain summertime comfort this might increase to 60 or 70%.

The glass is clear. No solar tinting was applied because it was felt that it would look odd if when the windows were open the occupants perceived a colour difference between the light through the solar tinted glass and daylight.

Options for additional solar control are mainly internal blinds, mid-pane blinds (see Figure 4.5) and external shades. Fixed external louvres have a principal disadvantage of an unavoidable reduction in passive solar gain and daylight. Depending on their design they may also be ineffective at many times of the year and may restrict the view out. For the EB, in spite of some reservations about maintenance, it was decided to use external motorized glass louvres (Figure 14.8), in part because of their important experimental value; the louvres are under BMS control normally and the occupants can override the settings at any time with the hand-held controllers. The general approach was seen as a possible way of avoiding the 'blinds down, lights on' syndrome.

Instead of an opaque louvre which when closed would exclude all daylight, 10 mm toughened clear float glass with a white ceramic coating on the underside was used. This has a light transmission of 40% and a reflectance of 50%, and gives a reasonable balance among solar gain, daylight and glare control. As a further line of defence, on the south façade roller blinds can be used; these blinds are also fitted on the north façade to deal with glare.

The south façade, with its motorized louvres, is a prime example of 'butterfly' design (see Section 4.7). The louvres are controlled by solar position and external light level. If the Sun cannot get in the windows the louvres slope slightly towards the

14.8 External glass louvres.

building in what is known as the light-shelf mode. In that position some light is reflected from them into the building.

If the Sun can get in the windows the BMS sets the louvre position to block out direct solar radiation. This is provided the external light level is above its adjustable set point; below the set point the Sun is let in. Careful consideration was given to siting the lowest louvre. Its height of 1.8 m from finished floor level resulted from designing to allow clear views out and to ensure that reflection from its upper surface would not annoy people standing in the office.

Daylight levels have proved to be reasonable. Figure 14.9 shows some indicative measured results with the external louvres in the open position.

14.6 Lighting

Daylighting is complemented by a high-efficiency, automatically controlled lighting system consisting of:

1. luminaires with high-efficiency TL5 linear fluorescent tubes (output 104 lm/W) with high frequency dimmable control gear;
2. a controller which can turn lights on and off, or adjust their output according to internal light levels;
3. PIR (passive infrared) movement detectors. The normal default setting is lights off, but in the presence of people these detectors turn the lights on if required;

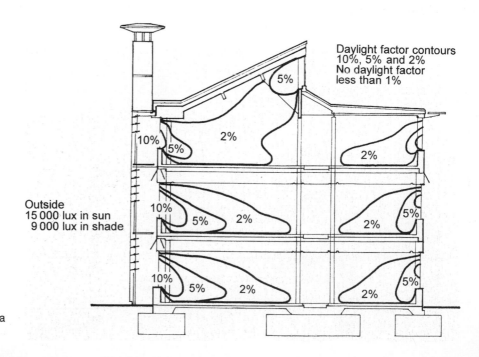

14.9 Daylight factor on a sunny day in winter.

4. hand-held infrared light controllers which allow users to override the pro-
    grammed settings, thus giving them control of their environment. These control-
    lers are integrated with the BMS and also permit control of the ventilation and
    overriding the position of the external louvres.

The offices are lit to a general light level of about 300 lux with fluorescent task
lights being provided to give a higher illuminance where needed. For the principal
linear luminaires the output is about 40% upwards (much of which is reflected back
down) and 60% downwards. This was done both for visual effect and to provide a
more uniform overall light level within the office. Total installed lighting load is 7.3
W/m^2. Obviously the use of high-efficiency lamps and a control system which provides light only in the quantity needed, and when needed, helped to meet the tight
electricity consumption design target of 36 kWh/m^2 yr. But there was also a significant 'hidden' advantage in that it is estimated that without this approach the internal
peak temperature during the day would be about 1 °C higher.

14.7 Photovoltaic panels

At an early stage of the project photovoltaic (PV) panels were examined but budget
restrictions ruled out their inclusion. Realizing that PVs were likely to be incorporated at a later date, the south-facing roof was designed so that panels could be
added. Then as the building approached completion the interest of the Department
of the Environment, Transport and the Regions led to an area on the west side of
the south façade (Figure 14.10) being used for thin film silicon (TFS) panels (see
Appendix D). Part of the PV cost was offset against the cost of cladding with another
(albeit cheaper) material.

14.10 PV panels on the
south façade.

At the EB, the gross façade area of PV panels in an aluminium frame is 46.9 m^2, with an active solar panel area of 38.6 m^2. This active solar panel area in test conditions is capable of a peak output of approximately 2 kW. In the actual installation on the south façade the maximum output is about 1.5 kW. The panels are arranged in two parallel banks of 16 and 18 panels. Each bank will provide a maximum of about 850 W of DC power at 96 volts. The power is transformed from DC to 230 volts AC by two inverters, which in turn feed power into the building's main electrical switchboard as shown in Figure 14.11. The inverters incorporate safety interlocks

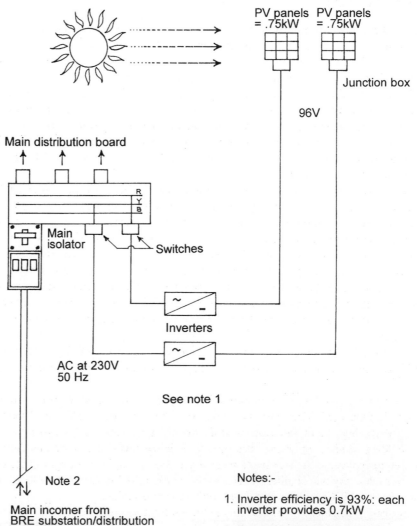

14.11 Basic electrical schematic with photovoltaics.

Main distribution board

Main isolator

Switches

Inverters

AC at 230V 50 Hz

See note 1

PV panels = .75kW   PV panels = .75kW

Junction box

96V

Note 2

Main incomer from BRE substation/distribution network

Notes:-

1. Inverter efficiency is 93%: each inverter provides 0.7kW

2. Excess energy is exported to the BRE site grid

which ensure that no AC power is exported from them if the switchboard they feed is off. There tends to be a good match (from Monday to Friday) between availability of PV generated electricity and demand in office buildings with their customary daytime occupancy. This is true at the EB also and so normally the generated power is used in the building itself. When supply exceeds demand the energy is exported to the rest of the BRE site. The lifetime of the installation is expected to be at least 25 years.

To put the PV contribution in perspective the maximum output of 1500 W will meet the small power requirement (i.e. socket outlets feeding computers and minor electrical items) for about 100 m^2 of the EB offices (at 15 W/m^2).

14.8 Materials

The EB provided an opportunity to examine what the construction industry could do if more attention were paid to sustainability (see Section 6.3).

Firstly, the existing building on the site had to be demolished and every attempt was made to recycle as much of the material as possible – this amounted to 96% by volume, with the remaining 4% going to landfill. Brickwork and concrete were crushed and used as hardcore under the new building. Timber was sent to a firm in Cornwall which specializes in making furniture with recycled wood. Steel roof trusses were cut up and sent off for melting down and recycling. The electrical accessories and items of mechanical plant were given to a charity which uses such equipment for community projects.

Secondly, in constructing the new building, careful consideration was given to the choice of materials. Thus, the renewable resource of a timber ceiling was used on the top floor as a compromise between a structural criterion to keep the weight low and an environmental one to keep the thermal mass up. The roof itself is aluminium sheet on the basis that aluminium is an easily recyclable material. Recycled aggregates replaced coarse aggregate in over 1500 m^3 of concrete for foundations, floor slabs, structural columns and intermediate floors. The recycled aggregates came from a building in London that was being demolished and crushing occurred at a plant in the London area; thus transportation costs associated with recycling were reduced.

In place of 100% cement in the concrete, a mixture of 40% cement and 60% ground granulated blast furnace slag, a by-product of the steel industry, was used.

To clad the building approximately 80 000 yellow London stock bricks over 100 years old were used. Their visual contrast with the gleaming stainless steel stack tops and the glazing strikes an attractive and symbolic note for future construction.

Building services, of course, also involve major material considerations. Some are related to avoidance and substitution strategies. For example, by designing the building in such a way that significant cooling was not required it was possible to avoid the HCFCs and HFCs usually associated with air-conditioning.

Similarly, the choice of smaller, more efficient light fittings meant that less mercury was used in manufacturing the lamps, and less released by the power stations providing the energy for them.

14.9 Evaluation and performance

The first hurdle was the BREEAM evaluation of the building which resulted in a classification of 'excellent' (39 credits out of a possible 42 were achieved) as shown in Table 14.1.

The next test was the first period of summer occupancy. This was the exceptionally warm summer of 1997 (when, for example, the mean temperature in August was about 3 °C higher than the long-term mean of 16.5 °C), which demonstrated that the building met the brief. Figure 14.12 shows an analysis of internal temperatures on all three floors.

Table 14.1 BREEAM evaluation

Item	Credits	Notes
Carbon dioxide production due to energy consumption	9 out of 10	This building produces the lowest amount of carbon dioxide of any building so far surveyed under the BREEAM scheme.
Acid rain	1 out of 1	
Ozone depletion due to CFCs, HCFCs and halons	7 out of 7	No refrigerants used, CFC free insulation material.
Natural resources and recycled materials	4 out of 4	Major use of recycled materials in the building.
Storage of recyclable materials	1 out of 1	
Legionnaires' disease arising from wet cooling towers	1 out of 1	No cooling tower.
Local wind effects	1 out of 1	The building is of similar height to those which surround it.
Noise	1 out of 1	There are no nearby residential dwellings.
Overshadowing of other buildings and land	1 out of 1	The building replaces one of a similar size.
Water economy	1 out of 1	6-litre WC cisterns used.
Ecological value of the site	3 out of 3	Existing building land was used and it was improved upon.
Cyclists' facilities	0 out of 1	Available on the BRE site, but not extended under this contract.
Legionnaires' disease from domestic water systems	1 out of 1	Water system designed in accordance with CIBSE code TM13.
Ventilation, passive smoking and humidity	3 out of 3	Openable windows and naturally ventilated building.
Hazardous materials	2 out of 2	
Lighting	2 out of 2	High daylight level within the offices and energy efficient lighting installed.
Thermal comfort and overheating	1 out of 1	Detailed assessments carried out in accordance with CIBSE guide.
Indoor noise	0 out of 1	The M1 is too noisy at times to meet the noise criterion for the cellular offices with the windows open.

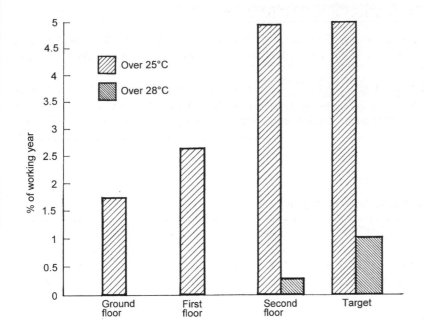

14.12 Internal summertime temperatures.

It is worth while noting that these are initial results when the systems were not completely free of faults and when only partial monitoring was taking place. The borehole cooling system was, in fact, not in use during the period because heat-metering equipment was being installed post-contract. Monitoring is ongoing and in several years a much more complete picture of performance will be available.

Although all floors meet the criteria of the brief one can also see a difference in performance between the top floor and the lower floors. It is likely that much of this is due to the lower thermal mass in the top floor ceiling.

Generally, peak temperatures in all offices are fairly steady and at least several degrees below external peak temperatures. It was a delight to visit the offices on one hot day in July 1997 and find one of the occupants wearing a pullover.

Energy use is also being monitored and results will be made available.

14.10 Future developments

How can one design an even more efficient building? First, we need to examine how energy is used in the EB. Figure 14.13 shows an approximate analysis of energy use and a development which we'll call the 'year 2010' version with a total energy consumption of only about 42 kWh/m^2 yr.

The reduction in energy consumption, which it should be noted could be achieved immediately, is due to factors such as those indicated in Table 14.2.

Now we can compare the total energy demand of about 42 kWh/m^2 yr with an output of, say, 100 kWh/m^2 yr from PV panels (see Appendix D). Over the course of

Table 14.2 Factors which reduce energy consumption

Item	Savings (kWh/m² yr)
Improved glazing (U-value: 1.1 W/m² K)	5
Insulated shutters on the windows	4
Mechanical ventilation system with heat recovery	2
Use of more energy-efficient computers in the offices	7

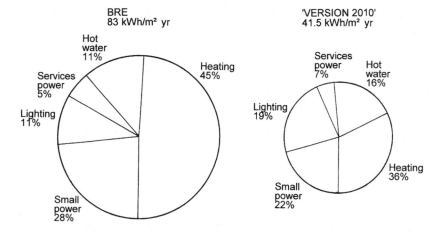

14.13 Energy use in the EB and in a 'year 2010' version.

a year 42 m² of PV panel will meet all the energy requirement of 100 m² of offices. Needless to say, this is encouraging. It takes us into the realm of what might be called Positive Energy Architecture – the architecture of those buildings that over the course of a year are net energy producers instead of being energy consumers. Their position relative to other buildings is shown in Figure 14.3. Developing these buildings will be one of the areas of cutting-edge design over the next 20 years. PVs will start to flourish (see Chapter 18) and in some cases will be joined by wind turbines (see Chapters 15 and 17), and progress will be made in materials selection and recycling.

14.11 Conclusion

The Environmental Building is an important step towards sustainable, positive energy buildings. Its most significant contribution is that it provides designers with a way of thinking about buildings. It isn't a kit of parts where architects select an external shade here or a stack there. Rather, it is an 'organism' in which elevation and section are equally important and equally necessary to provide fresh air, remove stale air and waste heat and maintain the comfort of the 'body'.

Project principals

Client	Building Research Establishment
Architect	Feilden Clegg Architects
Services engineer	Max Fordham LLP
Structural engineer	Buro Happold
Quantity surveyor	Turner & Townsend
Landscape architect	Nicholas Pearson Associates
Planning supervisor	Symonds Travers Morgan
Project manager	Bernard Williams Associates
Main contractor	John Sisk & Sons
Mechanical and electrical contractor	Norstead Mechanical & Electrical Engineers

References

1. Welsh, P.A. (1995) Testing the performance of terminals for ventilation systems, chimneys and flues. BRE Information Paper 5/95. BRE, Garston.
2. Jackson, M.T. and Linden, P.F. (1996) Flow Modelling Studies BRE Low Energy Office Building. Cambridge Environmental Research Consultants Ltd, Cambridge, UK.

Further reading

Anon. (1995) A performance specification for the Energy Efficient Office of the Future. General Information Report 30 BRECSU. BRE, Garston.

Davies, M., Jackaway, A. and Littler, J. (1997) *PLEA 97: 14th International Conference on Passive and Low Energy Architecture, Kushiro, Japan*, Vol. 2, pp. 21–6.

Dewey, E. (n.d.) Monitoring Protocol for Building 16 Environmental Office of the Future. BRE, Watford, UK. BRE, Garston.

Hobbs, G. and Collins, R. (1997) Demonstration of reuse and recycling of materials: BRE Energy Efficient Office of the Future. IP 3/97. BRE, Garston.

Hobson, S. (1997) *Living With The Future*. M + E Design. December 20–21.

Thomas, R. (1996) The design of an underfloor cooling system using borehole water for the new offices for the Building Research Establishment. *CIBSE Conference Proceedings*.

Thomas, R., Davidson, P. and Clegg, P. (1995) The new office for the Building Research Establishment. *CIBSE Conference Proceedings*.

Various (1997) Green Piece – BRE's prototype offices by Feilden Clegg. RIBA Profile. April.

Additional information

1.Information about the building is available from the following websites:

(a)www.archinet.co.uk/maxfordham
(b)www.bre.co.uk

2.A CD ROM on the building has been published by Building Services and distributed with their October 1997 journal. Selected parts are being made available at the BRE Website.

Acknowledgements

We would like to thank the other members of the design team, researchers at the BRE, and all those who have contributed to this project which is an exemplar of collaboration.

The Millennium Centre, Dagenham

15.1 Introduction

The Millennium Centre (shown in Figure 15.1) is for visitors to the Eastbrookend Country Park and Chase Nature Reserve, 80 ha of reclaimed former gravel pits at the eastern end of Greater London. The pits had been filled, in part, with rubble from Second World War bombsites in London. As the local mayor said, 'One hundred years ago the site was natural and now it has been restored but in between it's been to hell and back'. For this and many other reasons the project is a paradigm.

Exposed to both sun and wind the site offered an opportunity to further a number of concepts of environmental design. The client, the London Borough of Barking and Dagenham, was strongly committed to the environment and wanted a building of high architectural quality that would act as a showcase for sustainable development and construction. In many ways its agenda is similar to that of the Environmental Building (Chapter 14). The broad principles of sustainable design for the building were to use as little energy as possible in both construction and use and to cause the least possible disturbance to the environment, in the widest sense and over the lifetime of the building.

15.1 The Millennium Centre.

15.2 Construction and structure

Figure 15.2 summarizes some of the key features of energy conservation and sustainability.

The building faces south to make the most use of passive solar gain. Downstairs there are exhibition spaces and WCs – on the first floor there are offices and a viewing gallery. Wood is used extensively throughout. Externally, there is Douglas fir boarding and inside the roof is made of laminated timber beams.

Walls are constructed with masonite wood-fibre composite studs. External walls have 240 mm of blown cellulose fibre insulation made from recycled paper and the roof has 200 mm of the same material; the U-values of these elements are

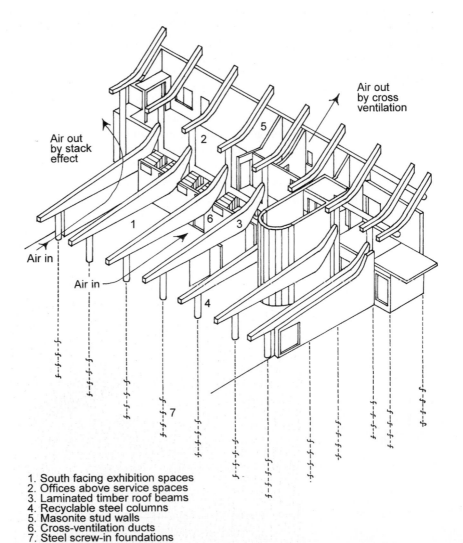

1. South facing exhibition spaces
2. Offices above service spaces
3. Laminated timber roof beams
4. Recyclable steel columns
5. Masonite stud walls
6. Cross-ventilation ducts
7. Steel screw-in foundations

15.2 Key design elements.

approximately 0.14 and 0.16 W/m^{2}K, respectively. Double glazing with low-emissivity coatings is used for the largest windows and triple glazing for the remaining windows.

A requirement for a robust ground floor surface suitable for muddy-booted park visitors led to the selection of concrete paving slabs sitting on 250 mm of compacted sand and gravel mix over hardboard protecting the DPM laid over 50 mm foamglass insulation. The paving slabs have a further advantage of contributing significant thermal mass to the building, thus lowering peak temperatures in the summer and in the winter storing heat so that internal temperatures do not fall too much. Services run in ducts in the sand/gravel mix.

A particularly novel feature of the building is the foundation which consists of 33 stainless steel screw anchors, each with a loading of approximately 85 kN, which go down about 10 m. Figure 15.3 shows a short length of typical anchor used for exhibition purposes just outside the building. A small concrete ring beam ties the tops of the anchors together and retains the built up dry floor construction. Steel columns are used where necessary to support longer spans of laminated timber beams.

The overall result is a building that at the end of its life can readily be taken apart and recycled.

15.3 Screw anchor exhibition display.

15.3 Environmental strategy and services

The need for heating in winter is kept low by insulating well, sealing the building and taking advantage of passive solar gain. Two gas-fired condensing boilers provide space and hot water heating.

In the summer, comfort is maintained by ensuring that the heat gains from the large south-facing glazing can be dispersed by ventilation and by using night cooling to lower the temperature of the whole building. The ventilation strategy (see Figure 15.2) allows for winds from several directions (by cross ventilation) and also for the less usual but more critical condition of low wind speeds (by stack-effect ventilation). In the summer during the day it is anticipated that much of the ground floor south-facing glazing will be open to welcome visitors; this will, of course, further aid ventilation. At night during the summer air paths of reduced size (for security reasons) are maintained and a small fan is brought on for night cooling if required.

Extensive natural daylighting is used; artificial lighting is all fluorescent.

The overall annual energy consumption of the building is expected to be approximately 50 kWh/m^2 for gas and 24 kWh/m^2 for electricity. Part of the electrical demand will be supplied by the wind turbine, visible on the right in Figure 15.1 and discussed below.

Photovoltaic panels were considered for the building but were too expensive for a rather tight budget. However, the slope of the large south-facing upper roof is at 45° and so when PV panels are hopefully installed at a later date their performance should not be far from optimal (see Appendix D).

15.4 The wind turbine

A three-bladed wind turbine (see Figure 15.4) with a swept diameter of 3.5 m sits on a 6.5 m high galvanized steel mast. Figure 15.5 shows the schematic of the system.

15.4 Wind turbine.

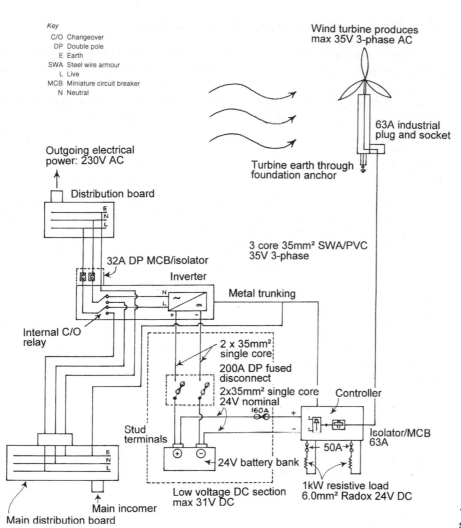

Key
C/O Changeover
DP Double pole
E Earth
SWA Steel wire armour
L Live
MCB Miniature circuit breaker
N Neutral

Wind turbine produces
max 35V 3-phase AC

63A industrial
plug and socket

Outgoing electrical
power: 230V AC

Turbine earth through
foundation anchor

Distribution board

3 core 35mm² SWA/PVC
35V 3-phase

32A DP MCB/isolator

Inverter

Metal trunking

Internal C/O
relay

2 x 35mm²
single core

200A DP fused
disconnect

2x35mm² single core
24V nominal

Controller

160A

Stud
terminals

Isolator/MCB
63A

50A

24V battery bank

1kW resistive load
6.0mm² Radox 24V DC

Low voltage DC section
max 31V DC

Main incomer

Main distribution board

15.5 Electrical
schematic (simplified).

The wind turbine has a cut-in speed of 2.5 m/s and a rated wind speed of 12 m/s at which it produces 2.5 kW. Noise can be an issue with wind turbines but the low rotor speed of the model selected of about 300 rpm keeps the noise between 45 dBA (at 5 m/s) and 60 dBA (at 20 m/s); these figures are measured at the base of the mast.

Three phase 35 V AC is produced by the turbine and fed to a controller which converts the electricity to DC, then to batteries, an inverter and on to a distribution board serving lighting circuits. If energy is not available from the wind turbine (or from the batteries) the board is fed in the normal way from the mains. Any excess energy is used in 'dump' loads for heating.

15.5 Water and waste recycling

Rain-water from the roof is collected and led to an underground tank for use for irrigation.

Reed beds and a number of other novel waste disposal systems were considered at an early stage, but as mains sewers run very close to the site it was thought more economic to connect to them and use the limited project budget for other purposes.

15.6 Conclusion

The Millennium Centre sets out to make the best use of the environmental resources of sun, wind and water. As such it is a building which points the way into the 21st century.

Project principals

Client	London Borough of Barking and Dagenham
Architect	Penoyre & Prasad
Services engineer	Max Fordham LLP
Structural engineer	Buro Happold
Quantity surveyor	Peter W Gittins & Associates
Main contractor	RG Carter London
Mechanical contractor	Ascot Heating Ltd
Electrical contractor	Estelle Electrical Contractors Ltd
Wind turbine supplier	Proven Engineering

Further reading

Spring, M. (1998) Self-sustaining showcase. *Building*, 30 January, 38–42.
Wild, D. (1998) Eastend ecology: Penoyre & Prasad in Dagenham. *Architecture Today*, **87**, 38–47.

The Bedales Theatre

16.1 Introduction

Bedales Theatre (see Figure 16.1), in the rolling countryside of Hampshire, has a green (i.e. unseasoned) oak structure with walls and roof of treated larch boards. The use of oak is symbolic of a centuries-old English tradition of building and an echo of the more recent oak timber-framed library at Bedales designed by Ernest Gimson in the 1920s.

  The theatre, with seating for up to 300, is essentially a pyramid on a plinth. Form, plan, choice of materials and services all speak of a straightforward, honest, humane approach. The overall height is 18 m and the total floor area 687 m^2 (including balconies, foyer, workshop, etc.).

16.2 An environmental approach

Starting with materials, wood (when grown and harvested suitably) is, of course, one of our most sustainable. The timber structure is strengthened at key points with stainless steel, a recyclable material.

  Low energy consumption was a specific goal and the architects responded with

16.1 Bedales Theatre.

well-insulated walls (155 mm of Rockwool insulation) and roof (150 mm of the same); the U-values of both constructions are approximately 0.2 W/m^{2}K.

Heating is from the school's district heating scheme with control of it and ventilation by a building management system (BMS). In conditions of partial occupancy during the heating season energy is conserved by running the fan in the 'lantern' in reverse (see Section 16.3) to force warm air down.

With regard to the electrical systems, where possible – for example in the workshop – low energy fluorescent lighting was used.

It is difficult to compare precisely the energy performance of a school theatre like this with other buildings, but for a 4-hour performance in the summer it has been estimated that at Bedales the energy consumption for ventilation would be 5 kWh compared with 190 kWh for a conventional, air-conditioned theatre.

Avoidance of air-conditioning, with its high energy consumption and its dependence on CFCs or more commonly HCFCs, was mandatory. Many conventional theatres are either too hot or air-conditioned – the challenge of providing a comfortable environment at Bedales is discussed below.

16.3 Ventilation

A slightly sloping site was a gift to the design team as it meant that air could easily be introduced into the building at below floor level. A second important advantage of the site was that noise is not a problem and it was agreed with the client that the occasional plane flying overhead would not be disruptive.

Theatres are characterized by high loadings from the (hopefully full) audiences and intermittently high loadings from stage lighting. Approximately 50% of the energy consumed by the stage lighting is a heat load to the occupied space.[1] In the case of Bedales, calculations were based on loadings of approximately 100 W/m^2 for people and 20 W/m^2 from lighting.

A goal of trying to ensure that the peak internal temperature did not exceed the peak external temperature by more than 3 °C was set. The key issue then became how to design the building to incorporate thermal mass where possible and to ensure that the ventilation paths were sufficient and worked with the mass.

Thermal mass was most easily incorporated in the plinth, which took the form of a concrete base with blockwork walls. The timber pyramidal structure above has only moderate mass (the roof admittance is 2.3 W/m^{2}K).

Figure 16.2 shows a schematic section through the theatre. Note the ample open areas for inlet and outlet air (5 and 6% of the floor area, respectively) and the high building height. The general approach to stack effect ventilation is similar to that used at De Montfort (Chapter 12) with the site and building form lending themselves perfectly to the task. On the other hand, the absence of high thermal mass in the soffit or roof is a significant difference.

Air enters at low level and passes underneath the seats. A striking illustration of the efficiency of the ventilation is seen in Figure 16.3 which shows the theatre seats with their pre-opening plastic covers. The stack effect has caused the covers to billow out. Air then rises through the theatre (see Figure 16.4) and out of the top.

Air flows straight through the building. Heat recovery would neither have been

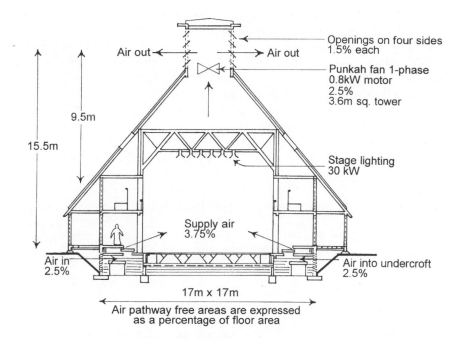

Openings on four sides
1.5% each

Air out ←   → Air out

Punkah fan 1-phase
0.8kW motor
2.5%
3.6m sq. tower

9.5m

15.5m

Stage lighting
30 kW

Supply air
3.75%

Air in
2.5%

Air into undercroft
2.5%

17m x 17m

Air pathway free areas are expressed
as a percentage of floor area

16.2 Section through
theatre.

16.3 Seats with plastic
covers before opening of
the theatre.

very practical nor economical, the main problem with occupied theatres being an
excess of heat rather than a deficit. A carbon dioxide sensor at high level controls flow
to ensure sufficient fresh air. If additional air is required for cooling, temperature
sensors cause motorized inlet dampers and outlet panels to open (providing that doing

16.4 Interior of theatre.

so will result in a lowering of the internal temperature – this is similar to a condition of the external temperature being lower than the internal, but not identical). A simple but elegant Punkah fan can be brought on to assist flow if required. Using this system the building can be cooled at night to lower its temperature in preparation for a performance the next day.

A less well recognized but important characteristic of the system is that there is easy and safe access to every component – every sensor, motor and the fan. This has helped both commissioning and maintenance.

Partial monitoring of conditions under a heat load test indicated that throughout almost all of the occupied zone temperatures are within the 3 °C target set. Temperatures would have been lower if it had proved possible to incorporate more thermal mass, thus emphasizing again the importance of this aspect of the design.

To our knowledge Bedales is the first theatre to have been specifically designed for assisted natural ventilation, with its modern use of controls and its integration with the building fabric and form. The engineering (both environmental and structural) complements the architecture and the result (visitors agree) is a delight.

Equally important and encouraging is that the basic principles of the design can be applied to theatres in urban situations by the addition of acoustic installation in the inlet and outlet air paths,[2] thus offering a new environmental approach to this and similar building types.

Project principals

Design consortium	Feilden Clegg Architects
	Roderick James Architect
	Carpenter Oak Woodland Company
Structural engineer	Ian Duncan
Services engineer	Max Fordham LLP
Client project manager	Sir Hugh Beach, Tim Battle
Client construction manager	Paul Buxey
Fire engineer	Buro Happold
Mechanical contractor	Corrall–Montenay
Electrical contractor	Southern Electrical Contracting

References

1. Quincey, R., Knowles, N. and Thomas, R. (1997) The design of assisted naturally ventilated theatres. *Proceedings of the 1997 CIBSE National Conference.*
2. Ibid.

Further reading

Evans, B. (1996) Theatre in the frame. *Architects' Journal*, **203**(6), 35–44.

CHAPTER 17

Beaufort Court

Beaufort Court in King's Langley, Hertfordshire is the new head office of Renewable Energy Systems (RES), one of the world's leading wind energy companies.

The site was formerly the Ovaltine Egg Farm, built in the 1930s in the Arts and Crafts style. The ground floor was used for housing up to 50,000 chickens, and for collecting and packing the eggs used in the production of the Ovaltine malted drink. The first floor was used for the storage of chicken feed and bedding material.

When RES purchased the site in 2000 it had been derelict for several years and the timber frame building was in relatively poor condition. The challenge for the design team was to turn the deteriorating farm buildings into a modern and sustainable office space complete with visitors' centre.

The brief included minimizing the energy consumption of the building and incorporating a range of renewable energy technologies into the site, which would be demonstrated to the public as part of the visitor facility.

17.1 Design constraints

Although the Egg Farm is not a listed building, it was considered to have historical significance, so one of the planning constraints was that the external appearance could not be substantially changed. This meant that any extension to the building had to be done within the central courtyard space, which would not affect the external view of the building.

As the development was within the Green Belt, the carpark and new ancillary building housing some of the renewable energy technologies had to be discreetly sunk so as to be hidden from view as much as possible.

Due to the location between the M25 motorway and the West Coast Mainline railway, noise was an important consideration. The noise level at the perimeter of the building was measured as 77 dB(A) when an Intercity train passed. For this reason, natural ventilation was felt to be inappropriate. All glazing on the external façade was sealed against noise. Openable windows and rooflights facing into the courtyard were used as they were less susceptible to noise.

17.2 Minimizing energy demand

The buildings are particularly well daylit, especially on the ground floor, so the need for artificial lighting is limited. On the first floor, rooflights have been introduced to bring daylight into the previously gloomy storage space.

17.1 Beaufort Court.

Office areas use fluorescent tube fittings with dimmable high frequency gear. A combined presence detector and photocell integrated into each fitting allows the electric lights to be on only when the room is occupied and when daylight levels are inadequate. As the internal partitioning of the office spaces is flexible, the lighting controls are programmable such that luminaires in the same room can be switched together. The individual photocells allow each luminaire to dim by an appropriate amount to meet a minimum design level of 300 lux, such that the luminaires closest to the windows are dimmed more than those at the back of the room.

As a low energy alternative to air conditioning, ground water is used to cool the building in the summer. This is discussed in more detail in section 17.3. To allow the ground water cooling system to achieve comfortable internal temperatures, the cooling load was minimised via solar shading, including fixed external shades and vegetation. Hornbeam trees, shown in Figure 17.2, are located close to the external perimeter of the building on the south and south-west sides. In time, the branches of neighbouring trees will be trained to grow together to form a high level hedge, blocking the high summer sun but allowing views out through the trunks below. The use of a deciduous species, which loses its leaves in winter, allows the building to benefit from the heat of the low winter sun.

To minimize energy associated with the construction process, demolition waste was kept on site and reused as far as possible. For example, old concrete was crushed on site and used as hard-core under the building and in paths; some of the waste timber was retained for use in the biomass boiler. Reclaimed railway sleepers were used to make external steps. All the soil was retained on site and incorporated into the new landscape scheme.

17.2 Boxed hornbeams provide solar shading.

17.3 Renewable energy strategy

As the project progressed, the brief was extended to developing the site to be carbon neutral, that is to say that over the course of a year the carbon emissions associated with the heating, cooling and electrical power used on-site would be offset by an equivalent carbon saving from on-site renewable energy generation.

As wind energy is the core business of RES, they were keen to have a wind turbine on the site. The location meant that this would be seen by a large number of people and could act as a positive advertisement for wind power. Through their contacts within the industry, RES were able to source a second hand turbine from Holland which could provide all the electrical power to the site, with any excess exported to

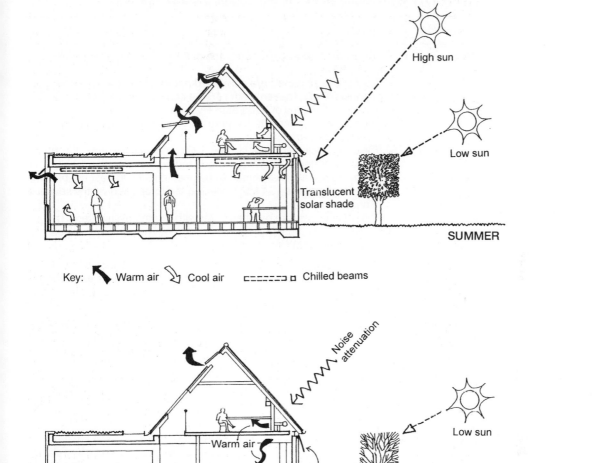

High sun

Low sun

Translucent
solar shade

SUMMER

Key: ◤ Warm air ◢ Cool air ⊏┈┈┈┈┐▫ Chilled beams

Noise
attenuation

Low sun

Warm air

Translucent
solar shade

Boxed hornbeam
trees

Sealed
glazing

WINTER

17.3 Winter and
summer thermal control
strategy.

the National Grid. Technical data relating to the wind turbine and other on-site renewables is given in Table 17.1.

The design team was already part of a European consortium funded by the EU to develop a combined photovoltaic and solar thermal panel (PVT) and to integrate it into buildings. A previous site had been identified for the pilot project, but due to timing this was transferred to Beaufort Court. Solar thermal panels are generally used to heat domestic hot water, but an objective of the EC project was that the solar thermal energy be used to heat the building. This presented an immediate challenge: the availability of solar energy and the heating needs of the building were about

Table 17.1 Renewable energy technologies used on the site

Technology	Details
Solar array	There are 22 solar collectors each of 6.88 m² giving a total area 151.36 m².
	7 out of the 22 panels are combined photovoltaic and solar thermal panels (PVT). The remaining 15 are conventional solar thermal panels. The area of integrated PV cells is 40 m² (net).
	The panels are connected in parallel. The array is south facing, with a tilt of 30 degrees.
	The thermal elements are by Zen Solar. The thermal peak power is approximately 100 kW. The predicted annual output is 40 MWh, of which 16 MWh is expected to be used directly to heat the building and the remaining 24 MWh is expected to be used via the seasonal heat store.
	The PV elements are by Shell Solar with Sunnyboy inverters. The nominal electrical peak power is 5.2 kW. The predicted annual output is 3.2 MWh.
Wind turbine	The turbine is rated at 225 kW. It is a second hand Vestas V29 with a hub height of 36 m and a rotor diameter of 29 m. It is expected to deliver 250 MWh a year, compared to the target building electrical demand of 115 MWh a year.
Biomass	The area planted is 5 hectares (50,000 m²). The expected yield is 60 oven dried tonnes per year.
	Miscanthus has a net calorific value of 17 MJ/kg[1] on a dry basis, equating to 280 MWh a year.
	The 100 kW boiler is made by Talbott. With a boiler efficiency of 80–85%, 220–240 MWh per year can be delivered to the building, more than meeting the predicted 200 MWh annual heating demand.
Seasonal heat store	The store consists of approximately 1000 m³ of water in the ground. The excavated hole is 6 m deep, tapering from a 20 m square at the surface to a 8 m square at base. It is insulated on top with 500 mm of polystyrene. The sides and base are not insulated. The liner is a Sarnafil plastic membrane.
	Over a 6 month period approximately 50% of the stored heat is expected to be lost.
	If the temperature of 1000 m³ of water is raised by 20 °C, 23 MWh of heat can be stored.
Borehole	The borehole is 75 m deep and 200 mm diameter. The pump delivers a maximum of 18 m³ per hour, and has variable speed control via an inverter drive. Water is extracted at 12 °C. With a 5 °C temperature rise, this can deliver 105 kW of cooling.
	The borehole is licensed for the extraction of up to 24,300 m³ per year, which can deliver up to 140 MWh of cooling.

6 months out of phase with each other. This resulted in the need to store the heat for some period of time. The development of a seasonal heat store was one of the focuses of the project.

Meeting the entire annual heating requirement with solar thermal energy was not practical and as the site had five hectares of agricultural land, the use of an energy crop for heating was a viable solution. Biomass consultants ADAS carried out a feasibility study on the field, advised on the choice of crop and carried out the planting. Miscanthus, also know as Elephant Grass, was selected over willow coppice as it gives better yields in sandy clay loam soil, is less susceptible to pests and disease, and can be harvested with conventional farm machinery whereas an alternative source such as willow, requires specialist harvesting equipment.

Miscanthus grows about three and a half metres high and can be harvested annually in January to March. It is usually baled for ease of transport and storage, and may be burnt either in whole bales (in a batch boiler) or after shredding, as is the case at Beaufort Court. The expected yield at the site is 60 oven dried tonnes per year, equivalent to 280 MWh (primary energy), which may be increased with irrigation to 100 tonnes per year.

The cooling system uses ground water extracted from a 75 m deep borehole which taps into the aquifer below the site. As the ground at this depth remains at a relatively constant temperature throughout the year, the ground may be used as a source of cooling in the summer, either via a heat pump or by means of direct water extraction.

The extracted water runs first through deep coils in the air handling units (AHUs) to cool and dehumidify the fresh air entering the buildings. Having been slightly warmed in the AHUs, the water then circulates through chilled-beams in the offices, consisting of finned copper pipes at high level. As the air flow past the chilled-beams is by natural convection, the heat transfer is not as great as in a forced air conditioning system. Also the temperature of the chilled-beams is limited to being above 15 °C to prevent condensation occurring. Together these factors limit the output of the system relative to conventional air conditioning and it is therefore necessary to minimize the cooling load, as described above, in order to achieve comfortable internal temperatures.

The electrical consumption of this cooling system is mostly due to the pumping energy of lifting a large mass of water up from the water table. The coefficient of performance has not yet been established but is expected to be around 10, making it about 3 times more energy efficient than a typical air conditioning system.

At the end of the cooling cycle, the ground water is used to irrigate part of the energy crop.

Figure 17.4 shows Beaufort Court's predicted annual loads as well as its predicted annual renewable energy generation.

17.4 Energy system integration

The wind turbine is connected to the site power distribution via a large power factor correction unit. The PV inverters are connected via a conventional distribution board. Both the turbine and the PV are metered separately so that their contributions can be monitored. The site is connected to the grid with an import-export meter. Any generated power in excess of the site's requirements is exported to the grid and sold to a green electricity supplier.

Since April 2002 all electricity suppliers have been obliged to provide a certain proportion of their electricity from renewable sources. Producers of renewable

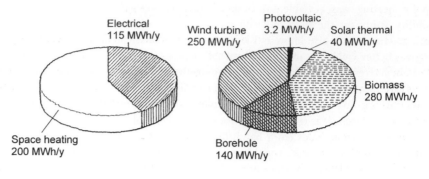

Electrical
115 MWh/y

Space heating
200 MWh/y

Wind turbine
250 MWh/y

Photovoltaic
3.2 MWh/y

Solar thermal
40 MWh/y

Biomass
280 MWh/y

Borehole
140 MWh/y

17.4 Annual energy predictions.

Predicted annual loads

Predicted annual renewable energy generation

electricity may obtain Renewable Obligations Certificates (ROCs) which can be sold to electricity suppliers, increasing the market value of the renewable electricity. RES are able to sell their wind and PV generated electricity under this scheme.

The heating operates on a two tier system. The relatively low grade heat from the solar thermal system (including that from the heat store) can only be used with a fairly low temperature heat sink, such as the external air. This heat is therefore used to preheat the incoming air to the building, providing up to 45% of the heating load at a given time. Air is distributed around the building in ducts at a seasonally compensated temperature (15–22 °C). Small heating coils on the outlet to each room provide individual room or zone control. The water in these coils is heated by the boilers, again to a seasonally compensated temperature (50–80 °C). For the majority of the year, the biomass boiler can meet the entire heating load, but in extremely cold weather gas condensing boilers will provide additional heat to meet the peak demand. They also provide back up in the event of a problem with the biomass system.

As the output of the solar thermal system cannot be guaranteed, each air handling unit also has a second heating coil supplied from the boilers which can supplement or take the place of the solar coil when solar heat is unavailable.

The energy systems are monitored via the building management system (BMS) which logs temperatures and heat flows at regular intervals. A number of heat meters are included in the systems for this purpose. They incorporate a water flow meter (based on a turbine), temperature sensors in the flow and return pipes and an electronic integrator. Electrical energy consumption and generation are also monitored.

The monitoring has been particularly beneficial during the first year of operation for diagnosing system and controls problems, for optimising the performance of the systems and for identifying areas of energy consumption which may be reduced.

17.5 Energy targets and initial performance

The building has been occupied by RES since November 2003. Over the first year of operation, all site electrical loads were met by on-site renewable generation (wind and PV). Commissioning of the seasonal heat store was completed late summer 2004 and the biomass boiler was installed at the end of 2004. At the date of publication, the heating aspect of the carbon neutral target had not yet been tested.

In the first year of operation, the annual heating consumption was double the predicted amount. The target was based on the *Energy Efficient Office of the Future* guide and was ambitious.[2] The increased heating consumption is due to a number of factors. Visual inspection and the use of a smoke puffer has identified cracks in the building envelope leading to air leakage. The use of an infra-red camera has identified cold bridges compromising the insulation. These fabric issues are being resolved. The heating controls took over a year to commission adequately which increased energy consumption.

There are also a number of factors which are outside the direct control of the design and construction team, relating to how the building is used. The operating hours and the internal temperature set points were higher than anticipated and can account for increases in energy consumption. We anticipate the heating consumption for the second year of operation being closer to the target value.

The electrical consumption in the first year has exceeded the target by 65%. Two reasons for this are that the operating hours have been longer than anticipated and equipment is not always turned off at night.

To conclude, achieving a building with low energy consumption requires attention to detail in both design and construction, and an understanding by the occupants of how the building works. We continue to work closely with RES to optimize the performance of this exemplar building, which has proved inspirational both to those who work in it and to RES' numerous visitors and clients.

Project principals

Client	Renewable Energy Systems
Architects	Studio E Architects
Services engineers	Max Fordham LLP
Structural engineers	Dewhurst MacFarlane & Partners
Quantity surveyors	A S Friend & Partners
Project manager	King Sturge
Main contractor	Willmott Dixon
Mechanical and electrical contractor	Sotham Engineering

References

1. Jones, M.B. and Walsh, M. (2001) *Miscanthus for Energy and Fibre*, James & James, London.

2.   Anon. (1995) *A performance specification for the energy efficient office of the future*, BRE, Carbon Trust, London.

Further information

ADAS, now BioRenewables Ltd, www.bio–renewables.co.uk/home
Talbott Heating, www.talbotts.co.uk
Further details of the project and monitoring data can be found at
    www.beaufortcourt.com

Heelis, Central Office for the National Trust

The new central office for the National Trust, Heelis (shown on the front cover) is a highly innovative building that utilizes natural ventilation and natural daylight along with other sustainable features to achieve a BREEAM rating of 'Excellent' with an anticipated energy use of about 10% less than the Building Research Establishment's (BRE) Environmental Building (Chapter 14).

The building is located on the site of the Great Western Railway Works in Swindon, much of which was originally designed by Isambard Kingdom Brunel. It is an office of around 7000 m² for 470 people and also contains a public café, shop and membership recruitment area.

The brief from the National Trust was that the building should be 'an appropriate 21st century response to a highly significant historic site and that the industrial heritage should be in some way legible in the design of the new building'. Sustainability is a key issue in the National Trust's philosophy and therefore the building needed to achieve high quality benchmarks in aspects of sustainable design. It was also desirable that the building should be open plan to ensure good links between departments.

18.1 Building form

Two, three and four storey building options were investigated with the design team settling on a two-storey, deep-plan building which reflected the existing historic structures by Brunel and his successors.

The north-west and north-east elevations were orientated in response to the existing site layout, but the main public elevation, internal layout and roof were all orientated to face due north–south to control solar gain and daylighting. The roof was pitched at a 30° angle to allow the maximum potential output from the attached photovoltaic panels (PVs). Rooflights were placed on the north-facing side of the pitched roof, and the PVs were cantilevered off the south side to provide shading to the rooflights and prevent direct sunlight from entering during the summer, so limiting high solar gains (Figure 18.2). The first floor contains a number of voids in the office areas to allow daylight from the rooflights down to the ground floor and two courtyards to allow ventilation into the centre of the deep plan (Figure 18.1).

Given the deep plan nature of the building, its roof is the main interface between the internal and external environments. Formed from 80 mm thick exposed pre-cast concrete panels laid onto the internal steel structure with high levels of insulation above and an aluminium standing seam external finish, the roof provides thermal mass for summer night-time cooling. The opening vents in the roof were formed from

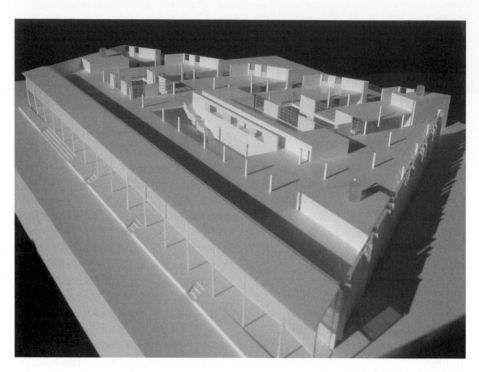

18.1 Model with roof removed exposing voids and courtyards.

lightweight insulated aluminium panels rather than glazed units to reduce the loads on the actuators.

18.2 Daylighting

As we have seen earlier, artificial lighting can represent a significant proportion of the running costs of a building (Chapter 3). Therefore, one of the main drivers in the design of the building was the desire to maximize the use of natural daylight.

External light levels of 6000 lux are exceeded for more than 80% of working hours in southern England (see Appendix A, Table A.3). Therefore, to obtain an internal daylight level of at least 300 lux for most of the working year (the minimum level recommended by CIBSE for office spaces),[1] an average daylight factor of 5% or more would be required.

BS8206 Lighting for Buildings Part 2, Code of Practice for Daylighting[2] states that if a predominantly daylit appearance is required and artificial lighting is expected to be used during the day, the average daylight factor should be not less than 2%; furthermore, if artificial lighting is not planned to be used during the daytime, an average daylight factor of 5% or more is required (as indicated above).

The two storey building form chosen has a number of first floor mezzanines within the main open plan office spaces. The plot ratio was about 1.5:1 so that two-thirds of the office could be directly roof lit and the remaining third could be generously lit under the mezzanine floors as shown in Figure 18.2. The most even light distribution to the ground floor is achieved with the rooflights running at right-angles to the mezzanines and voids, since the rooflights are north facing only.

A 1:5 scale mode of a typical area of the building was tested in an artificial sky. This achieved daylight factors of 15% on the first floor and 5% on the ground floor; varying as shown in Figure 18.2. These results put the building well within the recommendations of BS8206.

All fluorescent lighting within the office is provided with both presence and photo-cell detection and with dimming ballasts. As a result, artificial lighting only comes on to ensure minimum light levels are met, keeping electrical energy used for lighting to an absolute minimum.

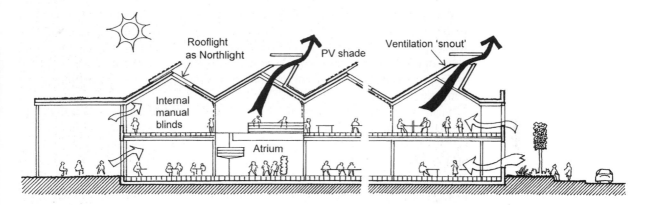

NORTH-SOUTH SECTION

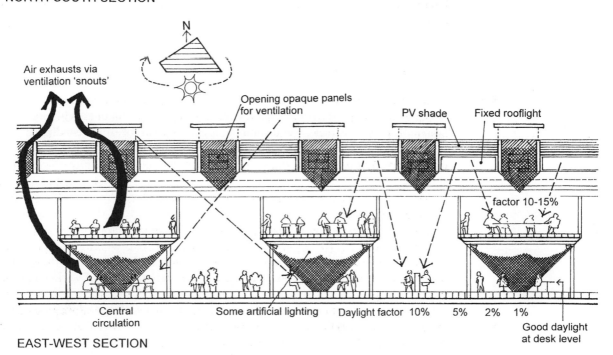

EAST-WEST SECTION

18.2 Sections indicating daylight and natural ventilation strategy.

18.3 Natural ventilation

It was decided at an early stage that in order to meet the Trust's sustainability brief, the building should be naturally ventilated if possible.

As we have seen previously in Chapter 9, in naturally ventilated buildings air moves because of pressure differences due to either the stack effect or wind effect, or a combination of both of these. There are occasional still periods but often the stack effect is the less pronounced of the two and it is common to use it as the design criterion.

Air is introduced via automatic opening windows and panels located around the perimeter of the building and the two courtyards which allow air into the centre of the deep plan layout. The windows are located at high level to enable them to be left open at night without compromising security. Air inlets along the south elevation and at discrete locations on the north elevation consist of large door-sized opening panels with external cast aluminium grilles for security.

Manual opening windows are also provided and all of the automatic openings have an over-ride facility to give the occupants a degree of control that can improve both actual and perceived comfort.

Air exhausts via openable roof lights that sit under large roof cowls which became affectionately known as 'snouts' due to their shape. These shelter the openings from rain and also allow air to exit regardless of wind direction. Propeller fans (see Figure 9.21) are installed in a few of the snouts to provide some mechanical back-up during still, hot periods.

The automatic openings are controlled by the building management system (BMS) to maintain comfortable temperatures within the building. During the summer these are achieved by ventilating at night to cool down the exposed concrete soffit of the roof and first floor slab. This thermal mass absorbs the internal gains during the day and emits them at night, thus effectively smoothing out the effects of fluctuations in external temperature to provide a more stable internal temperature.

Since the internal environments of naturally ventilated buildings rely on a combination of thermal mass, ventilation rates and external weather conditions it is not possible to dictate exact internal design conditions in the way that it is with air-conditioned buildings; rather the aim is to achieve reasonable conditions for most of the time.

The 'Energy Efficient Office of the Future' guide produced by the BRE on behalf of the government's DETR Best Practice Programme suggests that summer design conditions for a naturally ventilated building should be as follows:

Internal dry resultant temperature should not exceed 25 °C for more than 5% of working hours and should not exceed 28 °C for more than 1% of working hours.[3]

Dry resultant temperature is an average of air and room surface temperature (mean radiant temperature) calculated to produce an indication of 'occupant' temperature. This provides a better idea of comfort than air temperature alone as radiant heating or cooling from surfaces is taken into account. With a heavyweight building such as this and night-time ventilation, the radiant temperature becomes important in lowering the resultant temperature below the inside air temperature, since the exposed concrete surfaces will tend to be cooler than the air temperature for much of the day. Any air movement within the building will further reduce the 'occupant' temperature.

The performance of Heelis was assessed during the design stage by means of a computer thermal model. The weather data used in the model was that recorded at the BRE's offices in Watford during 1997. However, at the time of writing, the guidelines on the most suitable year to use for modelling are under review in light of recent successive years of hot weather and the predicted effects of global warming.

18.4 Winter ventilation

In a modern well-insulated and well-sealed building such as this, roughly half of heat loss occurring during winter is due to the ventilation required to control odour. A ventilation rate of around 8–10 l/s per person is required; the incoming fresh air needs to be heated from the outside temperature to the inside temperature of about 20 °C, and the exhaust air expelled to outside. A fully naturally ventilated building would do this via the external openings and thus lose warm air to outside.

By circulating the air through a heat exchanger however (and in the case of the Heelis this is a plate heat exchanger), the heat can be extracted from outgoing air before it is discharged outside and used to preheat the incoming air. In this way the ventilation heat loss is reduced by about 70%, less the losses from the fans that drive the air though the heat exchanger. This provides yet another way to reduce the overall running costs of the building. The pre-heated supply air is introduced into the building via the raised floor void and extracted from high level extract grilles (see Figure 18.3).

18.5 Cooling

The building is predominantly naturally ventilated, however mechanical cooling is provided where there are particularly high internal gains or stricter environmental conditions, as in several meeting rooms and the communications room which contains servers and other central IT equipment. Local fan coil units within the rooms are run on chilled water fed from two roof-mounted chillers. The chillers operate on Care 45, a hydrocarbon refrigerant, which is the HCFC 22 replacement with the lowest global warming potential (GWP) and also has a zero ozone depleting potential (ODP).

18.6 Sustainability

In order to clarify the sustainability agenda, a matrix was developed early on by the team which addressed areas such as operational energy use targets, on-site energy generation, ventilation, lighting controls, building materials, waste and water systems, bio-diversity, transportation and commissioning. The information was presented in

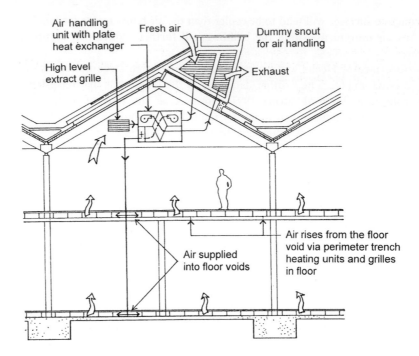

Air handling
unit with plate
heat exchanger

Fresh air

Dummy snout
for air handling

High level
extract grille

Exhaust

Air rises from the floor
void via perimeter trench
heating units and grilles
in floor

Air supplied
into floor voids

18.3 Winter-time
mechanical ventilation
strategy.

a table that defined targets under the headings of 'good practice', 'best practice', 'innovative' and 'pioneering'. It provided a concise way of determining how sustainable the building was likely to be, and where areas might be improved without substantial additional cost. It reflected many of the criteria within the BREEAM assessment and helped to guide the design to achieve an 'excellent' rating.

The use of PVC within the building was avoided by careful specification of materials including low smoke and fume cabling (LSF) and cast iron above-ground drainage. Waterless urinals were also installed to help reduce the overall water consumption.

To conclude, Heelis has taken daylighting and natural ventilation of offices in a new direction and demonstrates how a deep plan building can be made architecturally enticing, an outstanding place to work and highly sustainable with a very low running cost.

Project principals

Client	National Trust
Architects	Feilden Clegg Bradley LLP
Services engineers	Max Fordham LLP
Structural engineers	Adams Kara Taylor
Quantity surveyors	Davis Langdon
Main contractors	Moss Construction
Mechanical and electrical sub-contractors	IEI

References

1. Anon. (1993) *Lighting for Offices* (LG7:1993). CIBSE, London.
2. Anon. (1992) Lighting for Buildings: Part 2: Code of practice for daylighting. BS8206. British Standards Institute, London.
3. Anon. (1995) A performance specification for the energy efficient office of the future. BRE, Watford.

Appendices

Environmental data and the psychrometric chart

A.1 Solar data

Hours of sunshine

Figure A.1 and Table A.1 give data on sunshine in London.

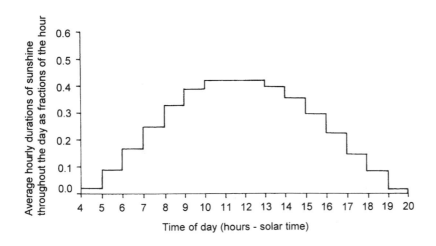

A.1 Hourly durations of sunshine for the whole year for London.[1]

Table A.1 Sunshine hours[2]

Period[a]	Average hours of sunshine per day in London
1. Spring 4 a.m.–8 p.m.	4.78
2. Summer 4 a.m.–8 p.m.	6.28
3. Autumn 6 a.m.–6 p.m.	3.51
4. Winter 7 a.m.–5 p.m.	1.71
5. Year	4.07

[a] Spring = March–May; Summer = June–August; Autumn = September–November; Winter = December–February.

Incident solar radiation in the United Kingdom

Figure A.2 gives solar radiation data.

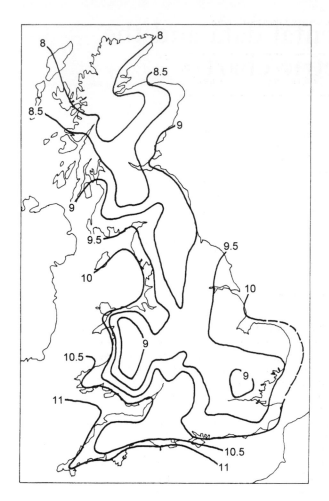

A.2 Annual mean of daily total global solar irradiation on a horizontal surface in MJ/m^2.^{3}

Solar altitudes and azimuths

Extensive data is available on solar position – Table A.2 gives a very basic introduction.

A.2 Daylighting

Table A.3 gives data on daylighting levels.

A.3 Temperature data

As for solar radiation, extensive temperature data is available. Table A.4 gives a very limited selection.

Table A.2 Approximate solar altitudes and azimuths at 52 °N[4]

Date and time	Altitude (deg.)	Azimuth (deg.)
December 21		
0900	5	139
1200	15	180
1500	5	221
March 21 and September 22		
0800	18	114
1200	38	180
1600	18	246
June 21		
0800	37	98
1200	62	180
1600	37	262

Table A.3 Illuminance at Kew, London[5]

Illuminance interval (lux)	Percentage of year[a] diffuse illuminance is above lower limit of range (%)	Percentage of year[a] total illuminance[b] is above lower limit of range (%)
0–1 000	100.0	100.0
1 000–5 000	94.3	94.4
5 000–25 000	84.1	84.6
25 000–50 000	30.7	48.6
50 000–75 000	1.3	23.8
75 000–100 000	0.0	8.0

[a] Based on a standard working year of 0900–1730, British Standard Time, April to October inclusive.
[b] Total illuminance is that due to light from the Sun and sky; diffuse illuminance is that due from the sky alone.

Table A.4 Daily mean dry bulb temperatures at Heathrow, London, 1976–95[7]

Month	Temperature (°C)	Month	Temperature (°C)
January	4.8	July	18.2
February	4.7	August	17.7
March	7.0	September	14.9
April	8.9	October	11.6
May	12.7	November	7.9
June	15.7	December	5.8

Average for period 1976–95 = 10.8 °C

Global warming is influencing every aspect of our lives, including, of course, building design. Over the last century the temperature of central England has risen by almost 1 °C and by the 2080s the average temperature of the UK may rise between 2 and 3.5 °C.[6] Peak temperatures in the summer are now reaching 37 °C and higher.

A fairly typical pattern for a hot summer's day is a minimum of 18 or 19 °C at night rising to say 28 °C or so at 3 or 4 p.m.

A.4 Wind

Table A.5 gives the Beaufort scale, which is useful in developing a personal sense of wind speeds.

Wind data is available from a number of sources.[8,9] The UK is one of the windiest countries in the world, and even in the southern parts of England the mean wind speeds are above 4–5 m/s 50% of the time, as can be seen in Figure A.3.

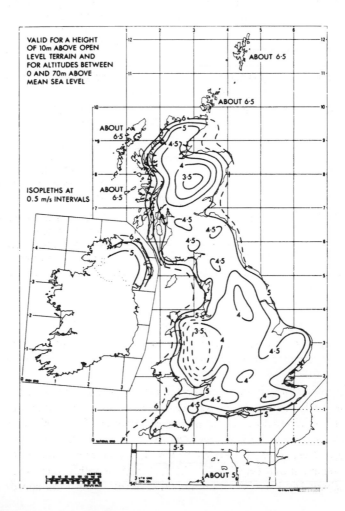

A.3 Isopleths of hourly mean wind speeds exceeded for 50% of the time over the UK.[11]

Table A.5 Beaufort scale

Beaufort number	Windspeed (m/s[a])	Description	Land condition	Comfort
0	0–0.5	Calm	Smoke rises vertically	No noticeable wind
1	0.5–1.5	Light air	Smoke drifts	
2	1.6–3.3	Light breeze	Leaves rustle	Wind felt on face
3	3.4–5.4	Gentle breeze	Wind extends flags	Hair disturbed, clothing flaps
4	5.5–7.9	Moderate breeze	Small branches in motion, raises dust and loose paper	Hair disarranged
5	8.0–10.7	Fresh breeze	Small trees in leaf begin to sway	Force of wind felt on body
6	10.8–13.8	Strong breeze	Whistling in telegraph wires, large branches in motion	Umbrellas used with difficulty. Difficult to walk steadily Noise in ears
7	13.9–17.1	Near gale	Whole trees in motion	Inconvenience in walking
8	17.2–20.7	Gale	Twigs broken from trees	Progress impeded. Balance difficult in gusts
9	20.8–24.4	Strong gale	Slight structural damage (chimney pots and slates)	People blown over in gusts
10	24.4–28.5	Storm	Seldom experienced inland. Trees up-rooted, considerable structural damage	

[a] Measured 10 m above sea or ground level.

Figure A.4 shows typical wind roses for Kew, London. The lengths of the lines are proportional to the amount of time the wind comes from the direction indicated.

An area of particular interest for designers of assisted naturally ventilated buildings is the frequency of low wind speeds and high temperatures, because these conditions are most likely to lead to discomfort. Little published analysis of available data appears to have been carried out. We have examined 10 years' of data from Kew, London[10] and found that June and July have fairly high mean temperatures (say 16–17 °C), high maximum temperatures (say 28–31 °C) and high monthly total solar radiation figures (say about 140 kWh/m^2 of horizontal surface). Mean wind speeds

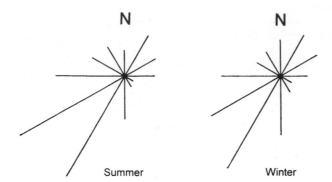

A.4 Wind roses for Kew, London.[12]

for June and July were 3.3 and 3.6 m/s, respectively, compared to a yearly mean of about 3.9 m/s. Examination of one particularly hot day in July with a minimum temperature of 15 °C at 4.00 a.m. and a peak of 30.6 °C at 4.00 p.m. showed a mean wind speed of 1.6 m/s with a minimum of 0.0 m/s and a peak of 4.9 m/s. The day in fact had a period of six continuous hours from midnight to 6.00 a.m. when the mean wind speed was less than 0.1 m/s; on the other hand, during the hottest period from noon to 6.00 p.m., the mean wind speed was 4.2 m/s. Obviously, much more analysis is needed. Cautious designers should probably plan on low wind speeds and high temperatures for short periods of interest, say, four hours. As the period of interest lengthens, it would appear reasonable to use somewhat higher wind speeds and lower temperatures.

A.5 Psychrometric chart

Figure A.5 shows the psychrometric chart, which relates temperature and moisture content.

The dry-bulb temperature is that measured by an ordinary thermometer. The wet-bulb temperature is measured by a thermometer with a wetted sleeve and so its readings are affected by the moisture content of the air (which is shown on the right-hand axis). The percentage saturation is the amount of moisture in the air at a given temperature compared to the amount of moisture in saturated air at the same temperature. Within the range of conditions normally encountered in buildings it can be taken as virtually the same value as the relative humidity. Relative humidity is defined as the ratio of the partial pressure of moist air at a given temperature to the partial pressure of the water vapour in saturated air at the same temperature.

As an example of the use of the chart, point P shows a condition of 20 °C dry bulb and 14 °C wet bulb, which is 50% saturation or, for our purposes, relative humidity. The moisture content of the air in this case is about 7 g/kg of dry air. The dewpoint, i.e. the point of 100% saturation, is just under 10 °C.

Table A.6 gives the frequency (on an annual basis) of percentage saturation ranges.

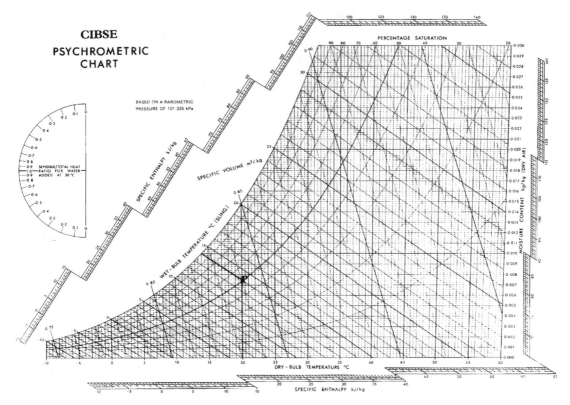

A.5 Psychrometric chart.[13]

Table A.6 Approximate annual distribution of percentage saturation for London (based on 20 years' data)[14]

Percentage saturation (%)	Percentage of year (%)
90–100	30
80–89	24
70–79	17
60–69	14
50–59	9
40–49	4
30–39	2

References

1. Ne'eman, E. and Light, W. (1975) Availability of sunshine. BRE CP 75/75. BRE, Garston.
2. Ibid., pp. 8–11.
3. Anon. (1986) *CIBSE Guide A2: Weather and Solar Data*, CIBSE, London.

4. Ibid., pp. A2–56.

5. Hunt, D.R.G. (1979) *Availability of Daylight*, BRE, Garston.

6. Anon. (2002) Climate Change Scenarios for the United Kingdom the UKCIP0Z Briefing Report. Centre, Norwich.

7. Anon. (2002) *CIBSE Guide J: Weather, Solar and Illuminance Data*, CIBSE, London.

8. Anon. (1999) *CIBSE Guide A: Environmental Design*, CIBSE, London.

9. Anon. (2001) *ASHRAE Handbook – Fundamentals: Weather Data*, ASHRAE, Atlanta.

10. Data from J. Littler of University of Westminster; analysis by R. Thomas, Max Fordham LLP.

11. See reference 3.

12. Anon. (1971) *IHVE Guide A2*, IHVE, London.

13. Anon. (n.d.) *Psychrometric Chart for Dry-bulb Temperatures: 10 °C to 60 °C*, CIBSE, London.

14. Holmes, M. and Adams, S. (1977) Coincidence of dry and wet bulb temperatures. TN 2/77. BSRIA, Bracknell. (Data derived from this source by the author.)

Calculation procedures

B.1 U-value calculations

The U-value is the rate of heat flow per unit area from the fluid (usually air) on the warm side of the element to the fluid (again usually air) on the cold side. The procedure for U-value calculations is given in numerous references[1,2,3] along with thermal conductivities for various materials. Values of thermal conductivity vary somewhat according to the reference consulted. Manufacturers will provide the most accurate test data for their products.

Here we shall merely give one example of a simplified calculation (e.g. wall ties are not accounted for) of a wall at Heelis, the National Trust Central Office (Chapter 18).

1. Construction
   – 15 mm dense gypsum plaster
   – 140 mm dense concrete block
   – 120 mm urethane insulation boards
   – 80 mm unventilated cavity
   – 103 mm brick
2. Internal surface resistance: use 0.13 m^2 K/W
   External surface resistance: use 0.04 m^2 K/W
   80 mm unventilated cavity: use 0.18 m^2 K/W

B.1 The wall under construction.

3. Thermal conductivity:
   dense Gypsum plaster: 0.57 W/mK
   dense concrete block: 1.13 W/mK
   urethane insulation board: 0.023 W/mK
   brick: 0.77 W/mK

4. To calculate the thermal resistance of a building element:

$$\text{Resistance} \frac{(\text{m}^2\,\text{K})}{\text{W}} = \frac{\text{Thickness (m)}}{\text{Conductivity (W/mK)}}$$

5. Add the resistance of all elements:

   Internal surface resistance $= 0.13\ \text{m}^2\,\text{K/W}$

   $$\text{Resistance of plaster} = \frac{0.015\ \text{m}}{0.57\ \text{W/mK}} = 0.03\ \text{m}^2\,\text{K/W}$$

   $$\text{Resistance of block} = \frac{0.14\ \text{m}}{1.13\ \text{W/mK}} = 0.12\ \text{m}^2\,\text{K/W}$$

   $$\text{Resistance of urethane} = \frac{0.12\ \text{m}}{0.023\ \text{W/mK}} = 5.22\ \text{m}^2\,\text{K/W}$$

   Resistance of cavity $= 0.18\ \text{m}^2\,\text{K/W}$

   $$\text{Resistance of brick} = \frac{0.103\ \text{m}}{0.77\ \text{W/mK}} = 0.13\ \text{m}^2\,\text{K/W}$$

   External surface resistance $= 0.04\ \text{m}^2\,\text{K/W}$

   Sum of resistances $= 5.85\ \text{m}^2\,\text{K/W}$

6. The U-value is the reciprocal of this:

$$\frac{1}{\text{Sum of resistances}} = \frac{1}{5.85\ \text{m}^2\,\text{K/W}} = 0.17\ \text{W/m}^2\,\text{K}$$

B.2 Daylighting calculations

A number of references deal with estimating daylight in buildings and the calculation of the average daylight factor (ADF).[4,5] The ADF (Chapter 5) for a side-lit interior is given by

$$\text{ADF} = \frac{TA_w\theta}{A(1-R^2)}$$

where T is the diffuse light transmittance of the glazing including the effects of dirt, blinds, obstructions and coverings; A_w is the window area (m^2); θ is the vertical angle subtended at the centre of the window by unobstructed sky; A is the total area of

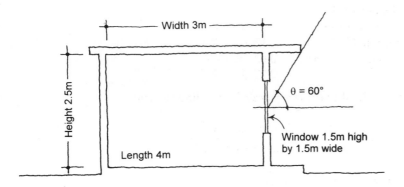

Width 3m — Height 2.5m — Length 4m — θ = 60° — Window 1.5m high by 1.5m wide

B.2 Daylighting for side-lit interiors.

indoor surfaces (ceiling, walls and floor, including glazing); and R is the area-weighted average reflectance of ceilings, walls and windows. We can apply this to the very simple case shown in Figure B.2.

Let us assume that $T = 0.75$ and

Reflectance of the ceiling	$= 0.7$
Reflectance of the wall	$= 0.5$
Reflectance of the window	$= 0.1$
Reflectance of the floor	$= 0.3$

The total area of the room is 59 m^2. The average reflectance is area weighted in the following way:

$(R$ side wall$) \times ($Area side wall$)$	$= (0.5)(3)(2.5)$	$= 3.75$
$(R$ side wall$) \times ($Area side wall$)$	$= (0.5)(3)(2.5)$	$= 3.75$
$(R$ back wall$) \times ($Area back wall$)$	$= (0.5)(4)(2.5)$	$= 5.00$
$(R$ front wall$) \times ($Area front wall$)$	$= (0.5)(10 - 2.25)$	$= 3.88$
$(R$ window$) \times ($Area window$)$	$= (0.1)(2.25)$	$= 0.23$
$(R$ ceiling$) \times ($Area ceiling$)$	$= (0.7)(4)(3)$	$= 8.4$
$(R$ floor$) \times ($Area floor$)$	$= (0.3)(4)(3)$	$= 3.6$
	Total	$= 28.61$

then,

$$R = \frac{28.61}{59} = 0.48$$

and

$$ADF = \frac{0.75(2.25)(60)}{59[1 - (0.48)(0.48)]} = 2.2\%$$

Thus, in conditions of a standard overcast sky of 5000 lux, the average light level would be $5000 \times 2.2\% = 110$ lux.

References

1. Anon. (1999) *CIBSE Guide A3: Thermal Properties of Building Structures*, CIBSE, London.
2. Burberry, R. (1992) *Environment and Services*, Longman, Harlow.
3. Anon. (2001) *ASHRAE Handbook – Fundamentals*, ASHRAE, Atlanta.
4. Anon. (1994) CIBSE code for interior lighting. CIBSE, London.
5. Anon. (1986) Estimating daylight in buildings: Part 2. BRE Digest 310. BRE, Garston.

Acoustics

In Chapter 2 we saw that sound levels are measured according to a logarithmic scale of decibels (dB) and that applying a frequency weighting to simulate the response of the human ear gives A-weighted decibel values (dB(A)). The decibel scale is a logarithmic scale defined relative to a suitable reference value.

The precise definition of the decibel scale depends on whether one is considering the sound power emitted by an acoustic source or the sound pressure in the air (or other medium) around a source. The definitions are given below.

Sound Power Level (L_W or PWL)
L_W (dB) = $10\log_{10}$ (Power/Reference Power)
Reference Power = 1×10^{-12} Watts

Sound Pressure Level (L_P or SPL)
L_P (dB) = $20\log_{10}$ (Pressure/Reference Pressure)
Reference Pressure = 2×10^{-5} Pascals

Sound power is the power emitted by an acoustic source, say a loud speaker in a room, and is measured in Watts. The power emitted by acoustic sources is very small in everyday terms – for example a human voice emits just one millionth of a Watt. The sound power emitted by the loud speaker can be converted to a corresponding Sound Power Level in dB.

The resulting loudness, or more precisely the Sound Pressure Level (Table C.1), in the room is dependent not only on the loudspeaker but also on the listener position

Table C.1 Representative sound pressure levels

Condition	Sound pressure level (dBA)
Threshold of hearing	10
Broadcasting studio	10–20
Living room in a quiet area at 7 a.m.	30
Typical business office	50–60
Listening to Chopin in a living room	50–65
Normal speech	55–65
Inside a train	55–70
Busy streets in urban area, e.g. Cambridge	70–75
Pop group at 20 m	100–110
Helicopter at 30 m	100–110
Threshold of pain	130

and on the characteristics of the room and its furnishings. The Sound Pressure Level associated with an unwanted source of sound is often termed a Noise Level.

Most noise sources are, of course, variable. To deal with this, noise levels can be examined over time and represented by suitable, single-figure statistical values. The average noise level in dB(A) over a time period T, or more properly the continuous equivalent noise level, is termed $L_{Aeq,T}$. $L_{A90,T}$ is the noise level in dB(A) exceeded for 90% of the time during time period T and is often used to describe the 'background' noise level. Similarly, $L_{A10,T}$ is the noise level in dB(A) exceeded for 10% of the time in time period T and is often used in the description of traffic noise.[1] The time period, T, is chosen according to the application, so, for example, for a school it could be the occupancy period.

The BREEAM evaluation procedure for new offices awards one credit if noise levels in large offices are between $L_{Aeq,T} = 45dB$ and $L_{Aeq,T} = 50dB$.[2]

It is also very common to see noise level recommendations given in terms of Noise Rating (NR) values. The approximate relationship between noise level in dB(A) and NR in dB is:[3]

dB(A) = NR + 6

Thus, if a recommendation for an air-conditioned conference room is NR25, the dB(A) equivalent should be about 31.

Fairly elaborate calculations are required to determine the noise level in a space.[4,5,6] The following relationship can be used as an approximate guide:

Noise Level in room (dB(A)) = Noise Level in adjacent space (dB(A)) − Sound Reduction Index of the separating construction (dB)

The Sound Reduction Index, R, of an element such as a wall, floor, door or window describes the proportion of incident sound that is not transmitted by that element.

$$R \text{ (dB)} = 10\log_{10} \text{(Incident Energy/Transmitted energy)}$$

Thus, if the noise level from a road is about 65 dB(A) and one is trying to achieve a noise level of 30dB(A) inside, the approximate façade sound reduction index required would be 35dB.

References

1. Anon. (1999) BS 8233 Sound Insulation and Noise Control for Buildings: British Standards Institute.
2. Anon. (2004) BREEAM/Offices/04. BRE, Garston.
3. Anon. (1986) *CIBSE Guide A1: Environmental Criteria for Design*, CIBSE, London.
4. Anon. (2000) BS EN 12354–1 Airborne Sound Insulation Between Rooms: British Standards Institute.
5. Anon. (2000) BS EN 12354–3 Airborne Sound Insulation Against Outdoor Sound: British Standards Institute.
6. Anon. (2002) *CIBSE Guide B5: Noise and Vibration Control for HVCA*, CIBSE, London.

Photovoltaics

Photovoltaic (PV) installations often consist of an array of PV modules and the associated wiring and control equipment. The modules are made up of cells. A typical crystalline cell might be 100 mm by 100 mm; module sizes might be 0.3 m^2 to 1 m^2 or larger.

Figure D.1 shows the spectral responsivity (i.e. the current generated per incident watt in standard conditions) for a representative monocrystalline silicon cell. Note that the cell responds to the visible range of 400–700 mm and some radiation outside it. Figure D.1 can be compared with Figure 2.4 showing the (broader) spectral composition of solar radiation.

A goal of PV design is to develop cells which make the maximum use of the solar radiation spectrum as economically as possible. PV efficiencies are increasing due to improved cell and module designs. In parallel the costs of components other than the PV modules have been reduced and project design and installation cuts have fallen as installers gain experience; generally, costs are being driven down by development of a

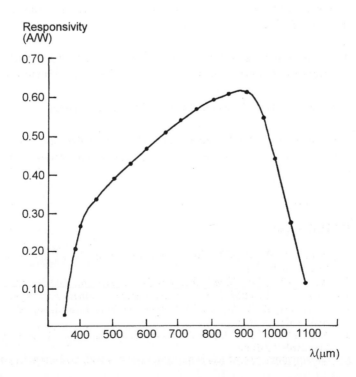

D.1 Spectral responsivity of a monocrystalline silicon cell.[7]

Table D.1 PV cell and module data[1]

Type	Approximate cell efficiency[a] (%)	Approximate module efficiency[a] (%)	Approximate installed cost (£/Wp)[f]	Construction
Monocrystalline silicon	13–17[b]	12–15[c]	10[g]	Single silicon crystal
Polycrystalline silicon	12–15[b]	11–14[c]	5.8–6.9[h]	Cast ingots of many crystals
Thin-film silicon	5[d]	4.5–4.9[d]	5.8[i]	Stack of series of p-n layers, so responsive to large band of wavelengths
Triple junction thin-film silicon	6–8	5–7	5.8–10[j]	Three amorphous layers with different band gaps
Heterojunction with intrinsic thin layer (HIT)	17–19[e]	14–16[e]	6.9–7.5[k]	Monocrystalline layer in a thin-film amorphous sandwich

[a] Efficiencies determined by testing under standard conditions for a solar irradiance of 1,000 W/m^2 with a spectrum corresponding to air mass 1.5 and a cell temperature of 25 °C.
[b] See reference 2.
[c] See reference 3.
[d] See reference 4.
[e] See reference 5.
[f] Note these are indicative installed cost for systems between 20 and 50 kWp based on south-facing arrays with 30° tilt. Costs have not been reduced for funds, grants, etc. Conversion based on mid market figures 11/02/05
[g] Building integrated monocrystalline SunSlates.
[h] Building mounted Sharp polycrystalline modules.
[i] Building mounted Kaneka amorphous silicon module.
[j] Building integrated Unisolar shingles and Unisolar solar metal roof.
[k] Building mounted Sanyo hybridcrystalline and amorphous modules.

mass market. Module costs in Europe have fallen by a modest 10% in the last 5 years, a small but steady decrease. In Japan, which leads the world in both the manufacture and installation of PVs, installed prices have fallen by approximately 80% in the last 10 years. Table D.1 displays PV cell and module data, it should be noted that these costs are indicative, with assumptions made on building mounting or integration techniques – installed costs contain a large degree of variation for this reason.

The basic technology is similar to that in the computer industry. The manufacture of monocrystalline solar cells starts with silicon dioxide, perhaps known best as the major constituent of sand, which is first reduced in arc furnaces to metallurgical grade silicon.[6] This is refined to semiconductor grade silicon in a polycrystalline form which is then melted. The Czochralski process is used to produce a large single crystal from the melt which is finely sliced into wafers. Further processing occurs and then a p-n junction ('p' for positive, 'n' for negative) is formed in the wafer by diffusing phosphorous into the silicon and introducing a small quantity of boron. Electrical metallic contacts are bonded to the wafer. Further processing occurs and the finished cells are then joined to form a PV module. The modules will often have a cover of low-iron glass which protects the front surface of the material while maintaining a high transmissivity. A structural frame is commonly used to protect the glass. Modules can then be formed into arrays of larger area.

Modules are tested in the standard conditions described above and their output is defined as the peak watts, thus a panel of 1 m^2 area in standard conditions of 1000 W/m^2 of radiation with an efficiency of 15% would have an output of 0.15 kWp. 1000 W/m^2 is a high level of solar radiation achieved in very sunny conditions; nonetheless, in London in clear sky conditions a south-facing wall at noon on 4 December receives about 650 W/m^2 and a south-facing surface tilted at 22.5° from the horizontal at noon on 21 June will receive about 945 W/m^2.

The actual output from an installation will almost always be lower than its peak output because of lower solar radiation levels, higher temperatures, less than optimal orientation, overshadowing, and so forth. As an example, we can consider the 39.5 kWp installation at the University of Northumbria, Newcastle upon Tyne (latitude 55 °N) (shown in Figure D.2). This is a 1960s concrete framed building in an urban situation partially shaded by other surrounding buildings and by a chimney in front of it. In 1994 it was reclad with PVs. The principle is rain screen cladding which provides protection from the elements while incorporating a ventilated space between cladding and building to equalize wind pressure and prevent condensation.[8] The PV arrays are tilted at 65° from the horizontal for a variety of considerations, including winter performance, window shading and aesthetics. The array consists of monocrystalline silicon modules with an efficiency of about 14%; it has achieved a maximum DC power output of 39 kW (in March 1995)[9] but most of the time its output is lower; the maximum daily output has been 166 kWh DC on a day in April 1995.

PV panels produce a DC current which can either be used directly or converted to AC (as at the Northumberland Building). If we simply concentrate on AC here, there are system losses in the cables leading to the DC to AC inverters and in the inverters themselves. A typical inverter efficiency over the course of a month might be about 90%. Commonly surplus energy is fed back to the national grid. An alternative for small or remote installations is to store power as DC in batteries.

Obviously, there are energy inputs into the manufacturing (and transportation) processes, but even in Newcastle, not normally known for its sunny clime, the energy payback period for the installation at the Northumberland Building was only 6.1 years; a similar analysis for a 1 MW installation in Toledo, Spain, gave a comparable period of 4.3 years.[10] At Heelis, the National Trust Central Headquarters (see Chapter 18), the expected energy payback period is approximately four years.[11]

Output will vary with both orientation and tilt as shown in Figure D.3.

Table D.2, developed during early work in the 1970s on solar heating systems,[12] gives radiation data for south-facing surfaces only but also includes horizontal ones.

Essentially it can be seen that it is preferable to remain within 20° or so of due south. With regard to tilt, there are several approaches – for grid connected systems, one will probably wish to maximize the annual yield. In this case, as is apparent from

D.2 Northumberland Building, University of Northumbria.

Table D.2 Incident solar radiation on a south-facing collector[a]

Tilt	Percentage of incident radiation relative to radiation on a surface with 30° tilt	Incident radiation (kWh/m² yr)
0° (horizontal)	92	910
30	100	993
45	99	981
60	92	917
90 (vertical)	70	695

[a] Average from Kew weather data, 1959–68.

Figure D.3, a tilt angle of 30° from the horizontal gives a better performance than one of 55°. A UK rule-of-thumb sometimes used for unobstructed collectors is that the tilt angle should equal the latitude minus 20°. Note, nonetheless, the somewhat surprisingly high figure for radiation on a horizontal surface and the potential it affords.

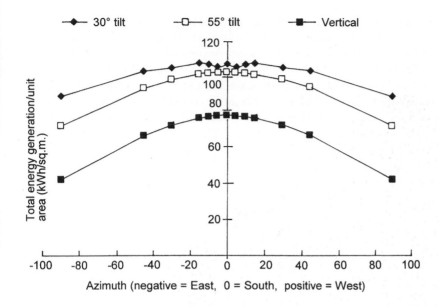

If collectors are obstructed, either by landforms or surrounding buildings, output will obviously fall and an assessment will need to be made of the probable energy loss. Similarly, banks of collectors arranged horizontally on a flat roof or on a wall may produce a degree of self-shading which will lower the annual output. Performance also varies with temperature – more so in the case of crystalline silicon cells than amorphous silicon. In sunny conditions cell temperatures of 75 °C are not unusual,[13] leading to impaired performance – for example, tests on crystalline modules have shown a drop in efficiency from about 15% at 25 °C to about 11% at 75 °C.[14] If the temperature can be reduced to 55 °C by ventilation the efficiency will be about 12 to 12.5%, and so the moral of the story is to ventilate behind the panels.

In reality, the range of orientation and tilt tends to be broader than a Platonic engineer might wish as the architect must reconcile conflicting considerations of the site context and the aesthetic appearance of the architectural forms. One encouraging development in PVs is that increased demand is leading to more choice and PV modules in colours other than a dark-blue or black are appearing. Since the other colours (for example, magenta or gold) are seen by causing some of the incident light to be reflected, the PV efficiency is about 20% lower than blue or black panels,[16] but this may be a price worth paying.

How do actual installations perform? At the Oxford Eco-house (latitude 51.8°C) the 4 kWp PV array is on the due-south-facing roof and is made up from 48 modules covering 30 m², the tilt angle is 40° from the horizontal. The annual yield is approximately 3000 kWh giving 100 kWh/m² yr.[17] The PVs are monocrystalline silicon and the array efficiency is 11%. The DC electricity provided by the array is converted to the AC by an inverter. At times there is an excess of energy and this can be sold back to the grid. Unfortunately a consumer cannot currently buy and sell at the same price. The electrical utilities will purchase at a maximum rate of 5p/kWh for a total of 70 kWh/month for every kilowatt peak installed. Once the export exceeds 70kWh/month/kWp the value dramatically decreases.[18] Economic costing is fraught with difficulties, but very roughly the cost of electricity from the installation is about 30 p/kWh (assuming a lifetime for the system of 30 years).[19] For high efficiency hybrid module installations with government funding this estimate could be reduced to 12 p/kWh (based on a DTI funding at time of writing of 50%).

At the Northumberland Building the economic issues are again not easy to resolve, but an estimate of £800/m² has been made as the cost of the PV system; this figure is over and above a conventional cladding system, which might cost £100/m² and includes the inverter and wiring and labour.[20] The annual output from the system is about 73 kWh/m²; using an array area of 285 m² the overall system efficiency is approximately 8%. Assuming a lifetime of 25 years and a discount rate of 0%, respectively, the unit cost is about 43 p/kWh;[21] higher estimates of the discount rate lead to higher unit costs. It is expected that reductions in the prices of PV laminates, cladding framework and electrical components will lead to lower costs for similar systems in the future.

PVs are clearly an idea whose time has come – but perhaps not quite yet. Nonetheless, the concept of a much cleaner source of power than we have had to date has captured the public imagination. As a result the PV market has seen a huge growth, with world installed capacity tripling in the last 5 years, and European capacity increasing four fold in the last decade.[22] Policy in the UK is following suit with DTI grants of £31,000,000 distributed from 2002 and new funding waves planned.

An approach that we might take is to follow many New York developers of a century or so ago who could see that their tall buildings would need lifts but who could not afford them. Immensely sensible, they incorporated the lift shaft during construction and added the lifts later. Can we do better (or even as well)?

References

1. Data from private communication with Solar Century, January 2005, unless otherwise stated.
2. Private communication, BP Solar, April 1998.
3. Anon. (1996) *Solar Electric – Building Homes with Solar Power*, Greenpeace, London.
4. Private communication, Intersolar, April 1998.
5. Anon. (2005) Sanyo North America Corporation Solar Products Website www.sanyo.com, January 2005.
6. Anon. (n.d.) Solar cells mono production. European Distributors Manual. BP Solar.
7. Courtesy of BP Solar, July 1997.
8. Anon. (1997) Architecturally integrated grid-connected PV façade at the University of Northumbria. ETSU S/P2/00171/REP. Contractor: Newcastle Photovoltaics Applications Centre, University of Northumbria.
9. Ibid.
10. Baumann, A.E., Ferguson, R.A.D. and Hill, R. (1997) External costs of the Toledo 1 MW PV plant and the Newcastle 40 kW BIPV façade. 14th EC PVSC Conference, Barcelona.
11. Data from Solar Century incorporated in Design Brief, New Central Office for the National Trust, Churchward, Swindon, October 2003.
12. Wozniak, S.J., (1979) *Solar Heating Systems for the UK*, HMSO, London
13. Brinkworth, B.J., Cross, B.M., Marshall, R.H. and Yang, W. (1997) Thermal regulation of photovoltaic cladding. *Solar energy*, **61**(3), 169–78.2.
14. Ibid.
15. Anon. (n.d.) Photovoltaics in buildings. BP Solar.

16. Roaf, S. and Walker, V. (eds) (n.d.) 21 AD Architectural Digest for the 21st Century, BP Solar/Oxford Brookes University, Oxford.
17. Ibid.
18. See reference 1.
19. See reference 16.
20. See reference 8.
21. See reference 8.
22. Anon. (2005) Solar Buzz Inc. www.solarbuzz.com, January 2005.

Further reading

Anon. (1997) Photovoltaic solar energy best practice stories. European Commission Directorate – General XVII for Energy THERMIE, Brussels.

Anon. (2004) *Planning and Installing Photovoltaic Systems – A Guide for Installers, Architects and Engineers*, The German Solar Energy Society, James and James, London.

Anon. (1998) Solar gain. *ECO*, February, 24–5.

Archer, M.D., Hill, R. (eds) (2001) *Clean Electricity from Photovoltaics*, Imperial College Press, London.

Lloyd-Jones, D., Hattersley, L., Agev, R., Koyama, A. (2000) *Photovoltaics in Buildings: BIPC Projects*, Crown, London.

Sick, F. and Erge, T. (1996) *Photovoltaics in Buildings. A Design Handbook for Architects and Engineers*, James and James, London.

Thomas, R. (ed.) (2001) *Photovoltaics and Architecture*, Spon Press, London.

Illustration acknowledgements

The author and publishers would like to thank the following individuals and organizations for permission to reproduce material. We have made every effort to contact and acknowledge copyright holders, but if any errors have been made we would be happy to correct them at a later printing.

Individuals and organizations

Architects Design Partnership: 9.24
ASHRAE, Atlanta: 7.4
John Barnard: 5.2
Paddy Boyle: 9.2
BP Solar: D.1, D.3
The British Wind Energy Association, London: 7.5
Martin Charles: 11.1
Chartered Institution of Building Services Engineers, London: A.3, A.5*
Cliché A. Attemand, L'Atelier du Regard, Orsay: 5.4
Combined Power Systems Ltd, Manchester: 7.2a
Conibear: 15.2
Peter Cook: 12.1, 12.6, 12.7
Edward Cullinan Architects: 8.3, 11.4
Department for Environment, Food and Rural Affairs: 10.4
Electricity Association Services Ltd, London: 9.22
Feilden Clegg Bradley Architects: 14.4, 14.5, 18.1
Max Fordham LLP: 1.3, 4.5, 12.4, 14.2, 14.6, 14.7, 14.8, 14.9, 14.10, 14.11, 14.12, 14.13, 15.3, 15.4, 15.5, 16.1, 16.2 (with Feilden Clegg Architects), 16.3, 17.2, B.1, D.2
Dennis Gilbert/VIEW: 13.1, 14.1, 14.4, 15.1
Short Ford & Associates: 12.2
Subinco: 4.6
Oak Taylor-Smith/Max Fordham LLP: Cover Photograph, 17.1
United Distillers, Edinburgh: 9.13
Harry Cory-Wright: 16.4

* Pads of charts are available from CIBSE, 222 Balham High Road, London SW12 9BS

Publications

Banham, R. (1969) *The Architecture of the Well-Tempered Environment*, The Architectural Press, London: 9.10

Clifton-Taylor, A. (1972) *The Pattern of English Building*, Faber & Faber, London: 1.2

Richards, J.M. (1958) *The Functional Tradition in Early Industrial Buildings*, The Architectural Press, London: 9.11, 9.12

INDEX

Note: Page numbers in **bold** refer to illustrations and those in *italics* refer to tables.

absorptance 11–12, *12*
absorption air-conditioning 137
Acheulean hut 3
acid rain 22
acoustics 18–21, 253–4
    see also noise; sound
    attenuation 46–9, **47**, 64, 127
    Charles Cryer Studio Theatre 183–4
    De Montfort University Queens
        Building 176–7
active noise controllers 48
active solar heating 88–9, **88**, **89**
ADF *see* average daylight factor
admittances 34–40, *38*, 50, 158
aerated materials, conductivity 9
agricultural waste 93, *93*
AHUs *see* air-handling units
air exhausts 234
air flows **121**, **122**, 234–5, **235**, **236**
air pollution 22–3
air quality 21–4, 64, 70
air tightness 45, 131, 197
air-conditioning 4, 31, 137–8
    BRE Environmental Building
        194–200
    controls 141
    energy flows **34**
    humidity control 24–5
air-handling units (AHUs) 160–1, 227
alligator farm, Florida 89, **89**
aluminium 9, 231–2
ambient temperature 7–8, 14
ammonia, natural refrigerants 69
Angers housing, France **58**
anthracite 84, *84*
    carbon dioxide emissions *85*
appliances 151
articulated forms 51, 98
artificial lighting 3–4, 103–7, *103*, *106*,
    174, 175–6, 232
asbestos 70
assisted natural ventilation 132–5,
    140–1
    Bedales Theatre 218–20, **219**

BRE Environmental Building 193–6,
    194, **195**, 197
atmospheric dust 22–3, *22*
atria 49, *97*, 101
attenuation, acoustic 46–9, **47**, 64, 127
Autarkic House, Cambridge University
    67
automatic control systems 50, 108,
    139–40, **141**
average daylight factor (ADF) 250–1

background noise 19–20, 47, 176
Beaufort Court 222–9, **223**, **224**, **225**
Beaufort scale 244, *245*
Bedales Theatre 217–21, **217**, **219**, **220**
BGS *see* British Geological Survey
biodiversity 56, 64
biogas 94
"biological age" 56
biomass, renewable energy 93, *226*,
    227, 229
blackbody 15
blinds 6, 44, **44**, 48–9, 101, 105, 201
BMS *see* building management systems
boilers
    combined heat and power 179–81
    gas-fired heating 110–14, **112**, **113**,
        197
    hot water service 148–9, 148–50, **150**
borehole cooling systems 184–8, *187*,
    191, 198–200, **198**, **199**, *226*, 227
BRE *see* Building Research
    Establishment
BREEAM *see* Building Research
    Establishment Environmental
    Assessment Method
British Geological Survey (BGS) 184
British Standards, Code of Practice for
    daylighting 232, 233
Broag gas-condensing boiler **112**
buffer spaces 49
building envelope 40–51
    see also buildings
    control 50

energy flows 31
form 50–2
heat loss 45
noise 46–9
second 'skins' 49
ventilation 45
building management systems (BMS)
    141–2, **142**
    Beaufort Court 228
    Bedales Theatre 218
    BRE Environment Building 195,
        196, 200
    De Montfort University Queens
        Building 178–9
    National Trust office 234
    RMC House 163
Building Regulations
    heat pumps 117
    heating demand of new buildings
        117
Building Research Establishment
    (BRE) 54, 60, 101, 125
    energy efficiency 234
    Environmental Building 190–209,
        190
    energy use 231
    evaluation 206–7
    heating 196–8, **198**
    light 200–3, **201**, **202**
    photovoltaics (PV) 203–5, **203**, **204**,
        208
    ventilation 193–6, **194**, **195**, 197
    wind distribution **191**
Building Research Establishment
    Environmental Assessment Method
    (BREEAM) 78, 190, 206–7, *206*, 254
    National Trust office BREEAM
        rating 231, 236
buildings
    see also building envelope
    embodied energy 70–3, **73**
    energy balances 29–35
    form 63, 231–2
    key materials 73–7

planning and design 37–53
renewable energy technologies 222,
224–9, *226*
spacing **57**, 58, **58**, **61**, 64

Calthorpe Park School, Hampshire 56
Cambridge University, Autarkic House
67
carbon dioxide
air quality 22, 23
boiler emissions 112–13
embodied energy 71, 72
emissions *85*
energy crops 93
energy use relationship *32*, 192, **192**
fossil fuels 29, 29–30, **29**, 84–5, *85*
global warming **16**, 17
heat pumps 87
hot water service systems 145–6,
146, 147
insulation 75
materials 67
mechanical ventilation 135, 136
natural refrigerants 69
carbon emissions reduction 117
carbon fuels 83
carbon monoxide 22, 23
carbon neutrality 224
cathedrals 61–2
cavity insulation 74, 74–6
CCT *see* correlated colour temperature
ceilings 73, 77, 98
cesspools 150
CFCs *see* chlorofluorocarbons
Charles Cryer Studio Theatre 183–8,
183
Chartered Institution of Building
Services Engineers (CIBSE)
internal daylight levels 232
psychrometric chart 246, **247**
chilled beams 227
chlorofluorocarbons (CFCs) 16, **16**, 67,
68–9, *69*, 87, 137
CHP *see* combined heat and power
Christ's College, Cambridge 136, **136**
CIBSE *see* Chartered Institution of
Building Services Engineers
CIE *see* Commission Internationale de
l'Eclairage
climate zones 54, **55**
Clivus Multrum dry toilets 150
co-generation installations *see*
combined heat and power
coal *82*, 83–4, *84*, *85*
cob cottages 3, **4**
coefficient of performance (COP),

heat pumps 87, 117
cold bridges 78
colour
radiation absorptions 40
reflectance 18
spectral response *17*
colour rendering index (CRI) 103
combined heat and power (CHP)
De Montfort University Queens
Building 179–81
fuel cell applications 94
hot water 149
installations 85–6, **85**
comfort 7–27
air quality 21–4
control relationship 7–8
light 17–18
moisture 24–5
sound 18–21
thermal 8–14, 110–44, 192, 234
Commission Internationale de
l'Eclairage (CIE) 103
compact forms 50–1
composite windows 47–9, **48**
computers
building management systems
141–2, **142**, 178
simulation 171, *172*, 193
concrete roof panels, National Trust
offices 231
concrete slabs 158, 161, 162, 194–6, **194**
condensation 13, 25, 45, 77
see also moisture
prevention 227
condensing boilers 111–14, **112**, **113**,
114, 197
conduction 9, **10**, 11
see also thermal conductivity
conservation, water 146
construction 77–8
Control of Noise at Work Regulations
19
controls
Beaufort Court 223, 228–30
BRE Environmental Building 196,
200
building envelope 50
De Montfort University Queens
Building 177–9
heating 139–40
hot water service 149
lighting 107–8, 233
ventilation 139–40, **140**
convection **10**, 11
conventional boilers 111, **112**, **113**, **114**
conversion factors xv-xvi

cooling systems
see also air conditioning; free cooling
energy reduction 31
groundwater
Beaufort Court *226*, 227
BRE Environmental Building 191,
198–200, **198**
Charles Cryer Studio Theatre
184–8, *184*, **185**, **186**, *187*
National Trust office 235
RMC House 158, 161, 162
types of 137–8
COP *see* coefficient of performance
correlated colour temperature (CCT)
103
costs
De Montfort University Queens
Building 179, *181*
photovoltaics 260
reduction of heating costs 235
courtyards 231, **232**
natural ventilation 234
CRI *see* colour rendering index
Crookham Church School **101**
cross ventilation 119, **121**, 123–4, 125
atria 49
BRE Environmental Building 195
De Montfort University Queens
Building 169, *170*
Millennium Centre **212**, 214
cross-flow heat exchanger 132, **133**
cupola 119

dampers 128, **128**
daylight 6, 96–102, 104, 105, 158
calculations 250–1, **251**
compact forms 50–1
control of 200–2, **201**
De Montfort University Queens
Building 60–1, **61**, 102, **102**,
173–4, **175**
factor 232–3, **233**
illuminance figures *243*
indicative daylight factors 96–100,
99, **100**
National Trust office 232–3
recommended daylight factors *96*
spectral composition **104**
sunny days **202**
DE *see* delivered energy
De Montfort University Queens
Building 37, 39, 77, 165–82, **165**, **177**
acoustics 176–7
axonometric **167**
building description 166–8
CHP installation 86, **86**

combined heat and power (CHP) unit 149, 179–81

daylight 60–1, **61**, 102, **102**, 173–4, **175**

insulation 74

lighting 169, 173–6, **174**

solar gain *168*

temperature 177–9, **179**

U-values *166*

ventilation 125, 136, 140, 168–73, *170*, *172*, **172**, **173**

decibels (dBA) 19, 253–4

deciduous trees 59, *60*

deforestation 68

delivered energy (DE) 81, *82*

demand histogram for hot water usage **146**

demolition waste 223

density *11*, *38*, *74*, *75*

Department for Education and Skills (DfES) 14

design constraints, Beaufort Court 222

destratification, water circulation systems 149

dewpoint 25, 246

DfES *see* Department for Education and Skills (DfES)

diffuse light levels **174**

dimmer switches 107

discomfort 14

displacement ventilation 137

Doha home 3, **5**

domestic coal, carbon dioxide emissions *85*

domestic heating *82*

double glazing 44, 76, **98**, 193, 212–13

double-leaf walls 21

draughts 128

drainage 76

cast iron above-ground 236

Sustainable Urban Drainage 56

dry toilets 150

ductwork ventilation 135–7

dust 22–3, *22*

eco-labelling 78

Ecology House at Stow, Massachusetts **56**

elastic modulus 73, *74*

electric heating 115–18, **117**

electricity 81, 82, 84–5

air heating systems 136

carbon dioxide emissions *85*

carbon dioxide production 117, 192

combined heat and power (CHP) 85–6

consumption, Beaufort Court 224–5, 229

energy conservation efficiency *82*

"high grade" energy 117–18

hot water service systems 147–8

power stations 85–6

renewable energy 117–18

fuel cells 93–4

hydroelectric power 92

photovoltaics 90, 203–5, **204**, 255

wind power 88, 90–2, *91*, **92**, 214–15, **214**, **215**

electrochromic glass 42

electromagnetic fields 70

electromagnetic spectrum 11, **11**, 15–17, **15**

elephant grass 227

embodied energy 70–3, 71, *72*, **73**, 75, *75*, 76

emittance 11–12, *12*

energy

see also electricity; fossil fuels; renewable energy; solar energy; solar radiation

admittances 38

balances 29–35, 41–2, **43**

Bedales Theatre 218

BRE Environmental Building 192–3, 207, **207**, *208*

conservation 151

consumption **29**, *32*, **32**, 222

controls 138–9

conversion efficiency 81, *82*

costs *82*

demand 222–3

efficiency 234

electromagnetic spectrum 11, **11**, 15–17, **15**

embodied 70–3, 71, *72*, **73**, 75, *75*, 76

flows 31–4, **34**

heating efficiency 114, 117, 139

hot water service systems 145, 146, *146*, 147

hydroelectric power 92

insulation 74

lighting systems 76, 107, 107–8

materials 70–3, *72*, 78

National Trust office 234, 235

renewable energy technologies 222, 224–9, **225**, *226*

renewable sources 87–94

RMC House 163, **163**

sources 81–94

stiffness 73, *74*

temperature control 8

units xv

wind power 88, 90–2, *91*, **92**, 214–5, **214**, **215**

windows 40–2, **41**

Environmental Building *see* Building Research Establishment (BRE) Environmental Building

environmental profiles 71

Europe, climate zones **55**

European Union

development funding 225

energy conservation 151, *151*

evaluation 78, 206–7

evaporation 7, **10**, 13

filter strips 56

filtration, assisted natural ventilation systems 134–5

finishes 77

fire precautions 178

flooding, reduction 56

floors 77

fluorescent lighting 76, 103–4, **103**, *106*

Beaufort Court 223

BRE Environmental Building 203

De Montfort University Queens Buildings 169, 175

National Trust offices 233

Fontevraud Abbey, France 57, **58**

forced convection 11

Forestry Stewardship Council (FSC) 68

form 37, 50–2, **62**

fossil fuels 3, 81, 82–5

carbon dioxide emissions 17, 29, 29–30, **29**, 84, *85*

energy conservation efficiency *82*

pollution 22

properties *84*

foundations 213

free cooling 161, 186

FSC *see* Forestry Stewardship Council

fuel cells, renewable energy 93–4

fuel oils 83, *85*

gas *82*, 83, *84*, *85*

air heating systems 136

boilers 110–14, **112**, **113**, **114**

hot water service systems 147–8

legislative requirements 116

gas oil 83, *84*

gaseous petroleum fuels *see* gas

geothermal energy 94

glare 6, 18, 31, **98**, 105

glass

electrochromic 42

solar radiation 40–2, *42*, 44, 48–9, **48**

spectral transmission curves **41**
thermochromic 42
glazing 76
  attenuation 46–9
  composite windows 47–9, **48**
  daylight factors 97, **99**, **100**
  energy use relationship 192–3
  solar gain 200–1, 214
  wind control 62
global warming 16, **16**, 17, 68, *69*, 84,
  117
  global warming potential (GWP) 235
greenhouse effect 68
greenhouse gases 16–17, **16**, 68
groundwater cooling systems 138
  Beaufort Court 223, *226*, 227
  BRE Environmental Building 191,
    198–200, **198**
  Charles Cryer Studio Theatre
    184–8, *184*, **185**, *186*, *187*
GWP *see* global warming

halons 68, *69*
HCFCs *see* hydrochlorofluorocarbons
HCs *see* hydrocarbons
health issues
  air pollution 23
  materials 70
  seasonal affective disorder 105
heat energy 81
  see also overheating; temperature
heat exchanger, air circulation 235
heat loss
  comfort factors 7–8, 9
  compact forms 50–1
  energy efficiency 32
  infiltration 77
  insulation 4, 9, 45
  rooflights 99–100
  U-values 41–2
  ventilation 235
  wind control 62
  windows 12
heat pipe heat exchanger 132, 134, **134**
heat pumps 86–7, 117
  coefficient of performance 87, 117
heat recovery 115, 135–6, 179, 197–8
  devices 132–4, **133**, **134**
  ducted air heating systems **136**, 137
heat store, seasonal 226–9, *226*
heat transfer 8–14
  admittances 37–40
  conduction 9, **10**, 11
  convection **10**, 11
  evaporation **10**, 13
  radiation **10**, 11–12

heating 110–18, **110**
  see also hot water services
  active solar 88–9, **88**, **89**
  Beaufort Court 225–9, *226*
  Bedales Theatre 218
  boilers 110–14
  BRE Environmental Building 196–8,
    198
  Building Regulations 117
  carbon dioxide emissions 29, 255
  combined heat and power (CHP)
    unit 179–80
  controls 139–40
  costs *82*, 235
  distribution systems/emitters
    114–18, **115**, **116**
  energy use **32**
  insulation 49
  RMC House 160, *163*
heavyweight buildings 38, 39
Heelis, National Trust central office *see*
  National Trust
HFCs *see* hydrofluorocarbons
high-efficiency boilers 111, 112, **113**,
  114, 197
highgrade fuels, lowgrade applications
  94
hornbeam trees 223, **224**
hot water service 145–50, **148**
  combined heat and power 179
  demand histogram **146**
  energy consumption *146*
  indicative decision tree **147**
  solar water heating 88–9, **88**, **89**
house mites 24
houses
  embodied energy 71–2, **73**
  energy consumption *32*
  energy flows **33**
  hot water service systems *146*
Housing Energy Efficiency Best
  Practice Programme 117
humidity 7, 24–5, 246
hydrocarbons 83
hydrocarbons (HCs), natural
  refrigerants 69
hydrochlorofluorocarbons (HCFCs)
  16, 68–9, *69*, 87, 137
hydroelectric power 92
hydrofluorocarbons (HFCs) 69, *69*, 235
hydrogen, fuel cells 93–4

illuminance xv, 96, 97, 103, *243*
indirect light 18
indoor pollutants 23–4
industrial heating *82*

Industrial Revolution 3
infiltration 131
infrared controllers 196, 202
insulation
  condensation 77
  energy consumption relationship 30
  glazing 49
  heat loss 4, 9, 45
  materials 69, 74–6, *75*, 77
  RMC House 158
  roof 231
  sound 20, 21, **21**, *42*, 176
  translucent 42
  walls 212
  windows 77
internal environments, naturally
  ventilated buildings 234
internal layout 50
interstitial condensation 25

kerosene 83, *84*
key materials 73–7
Kyoto protocol 117

landscaping 55, 56
Larkin Building 117, **117**
latent heat 7, 111
lead-based paint 70
life-cycle analyses 78
light *see* daylight; lighting; solar
  radiation
light shafts 102, **102**
lighting 96–109
  see also daylight
  artificial 3–4, 103–7, **103**, *106*, 174,
    175–6
  BRE Environmental Building 202–3
  comfort factors 17–18
  De Montfort University Queens
    Building 169, 173–6, **174**, **175**
  energy consumption 33
  glazing 49
  materials 76
  National Trust office 232–3
  recommended levels 97
  reflectance *98*
  RMC House 158, 161, *163*
  spectral power distribution **103**
lightshelves 101–2, **101**, 173
lightweight buildings 38, 39
liquefied natural gas (LNG) 83
liquid petrolem gas, carbon dioxide
  emissions *85*
liquid petroleum fuels 83, *84*
liquid petroleum gases (LPGs) 83, *84*
LNG *see* liquefied natural gas

London, sunshine hours *241*, **241**
longevity 72
'loose-fit' ventilation 195
Loughborough University, water
    conservation 145
louvres 126–9, **128**, 201–2, **201**
low smoke and fume cabling (LSF) 236
LPGs *see* liquid petroleum gases
LSF *see* low smoke and fume cabling
luminaires 76, 104, 107, 202, 203, 223

maltings 119, **119**
masonry walls 20
mass law **21**
materials 67–80
    absorptance/emittance *12*
    admittances 37–40, *38*
    BRE Environmental Building 206–7
    CFCs/HCFCs 68–9
    construction 77–8
    energy 70–3, *72*, **73**
    environmental aspects 67–8
    evaluation methods 78
    hcalth issucs 70
    insulation *75*
    key 73–7
    light reflectance *98*
    selection of 67
    stiffness 73, *74*
    thermal conductivity 9, *11*
    U-values *166*
mechanical ventilation 49, 135–7, 141,
    193
    Charles Cryer Studio Theatre
        186–8, *186*
    RMC House 158, 160–1, **160**, *163*
membrane sound absorbers 20
mercury lighting 175–6
methane **16**, 83, 150
methanol, fuel cells 93–4
mezzanines, open plan office spaces 232
micro-hydroelectric power 92
microclimate 55–6
Millenium Centre, Dagenham 211–16,
    211, **212**
miscanthus *226*, 227
mites 24
moisture 7, 24–5, 31, 45, 77
    see also condensation; humidity
    percentage saturation 246, *246*
    psychrometric chart 246, **247**
Montreal Protocol 68, *69*
mould 24, 25

National Rivers Authority (NRA) 184,
    199

National Trust central office, Heelis
    231–6, **232**, **233**, **236**, 249–51
    brief for building 231
    building form 231–2
    daylight calculations 250–1
    energy payback period 257
    U-values 249–50
natural gas 83, *84*
natural refrigerants 69–70
natural resources, water 146
natural ventilation 49, 51, 55, 62, 64,
    119–35
    checklist *127*
    controls 140–1
    costs *181*
    De Montfort University Queens
        Building **172**, 168–73, *170–1*, *172*,
        173
    National Trust office 231, **233**,
        234–5
    nitrogen oxides (NOx) **16**, 22–3, 84,
        113–14, 118
noise
    see also acoustics; sound
    acoustic louvres 127
    attenuation 46–9, **47**, 64, 127
    calculating sound levels 253–4, *253*
    comfort factors 18–21
    shelterbelts 64
Noise at Work Regulations 19
Northumberland Building, University
    of Northumbria 257, **258**, 260
Notre Dame de Paris 62
NRA *see* National Rivers Authority
nuclear power 82

occupants
    control 50
    heavyweight buildings 39–40
    infra-red controllers 196, 202
    satisfaction 163
    sick building syndrome 24
    thermal comfort 13–14
ODP *see* ozone depletion potential;
    ozone layer depletion
off-site fabrication of components 78
office, energy flows **34**
offices
    Beaufort Court 222–9, 256
    BRE Environmental Building
        190–209
    embodied energy 71
    energy use 35
    hot water systems *146*
    lighting levels *97*
    National Trust 231–6

RMC House 157–64
    temperature 14
    ventilation 125
offshore underwater tidal currents,
    renewable energy 94
oils *82*, 83, *84*, *85*
OPG *see* other petroleum gas
optimizers, heating systems 139–40
organic materials, conductivity 9
orientation 37, 56, 58, **59**, 231, 232–3,
    233
other petroleum gas (OPG), carbon
    dioxide emissions *85*
overheating 4, 6, 31, 38, 42, 49
Oxford Eco-House 259
oxygen 23, 64
ozone layer depletion 4, 16, 68–9,
    69
    ozone depletion potential (ODP) 68,
        235

paint 70
paraffin 83, *84*
partitions 21, 50
passive (PIR) movement detectors 202
PE *see* primary energy
percentage saturation 246, *247*
permeable surfaces 56
petroleum fuels 83
photocells, lighting controls 223
photoelectric cells *see* photovoltaics
photovoltaics (PV) 88, 90, 94, 255–60,
    255, *256*, *258*, **259**
    BRE Environmental Building 203–5,
        203, **204**, 208
    Millennium Centre 214
    panels 63
    photovoltaic and solar thermal panel
        (PVT) 225–6, *226*
    PV shade panels 231, **233**
pipework 76, 148–9
Planned Cost Effective Adaptation 78
planning
    building 37–53
    site 54–66
plastics 68, 75
    insulation materials 69
plate heat exchanger, air circulation 235
pollution
    air 22
    displacement ventilation 137
    fuel impurities 94
    global warming 16, **16**, 17, 68, *69*, 84
    indoor 23, 24
    lighting systems 76
    materials 67

ozone layer depletion 4, 16, 68–70, 69
   vegetation 64
   ventilation 118
ponds 56
porous sound absorbers 20
Positive Energy Architecture 208
power stations 85–6
presence detectors, lighting controls 223
pressure, units xv
primary energy (PE) 81, *82*
privacy, background noise 18–21
propellor fans 132, **132**
psychrometric chart 246, **247**
PVC 68, 76, 236
PVT *see* photovoltaics

Queens Building *see* De Montfort University Queens Building

radiation
   see also solar radiation
   electromagnetic 11, **11**, 15–17, **15**
   heat transfer **10**, 11–12
radioactivity 70
rain 55–6, 62
   rainwater harvesting systems 145
recycling 68, 72–3, 206, 216
reed beds, waste disposal 150
reflectance 18, *98*, 174, 250–1
refrigerants 67, 68–70, *69*, 87, 132, 137, 235
   as blowing agents 69
   natural 69–70
Regional Museum of Prehistory, Orgnac l'Aven 43
relative humidity 246
renewable electricity 228, 229
renewable energy sources 87–94, 117–18
   see also solar energy; wind
Renewable Energy Systems (RES) 222, 224, 229
renewable energy technologies 222, 224–9, **225**, *226*
Renewable Obligations Certificates (ROCs) 228
representative sound pressure levels *253*
RES *see* Renewable Energy Systems
reverberation time (RT) 20–1, 176
RMC House 39, 77, 157–64, **157**
   energy consumption 163, *163*
   internal conditions 162, **162**
   office area **158**, 159

site and building 162
Rockwool 74, 75, 176, 193
ROCs *see* Renewable Obligations Certificates
roof, thermal mass 231
rooflights 100–2, **100**, **101**, **102**, 158, 160
   National Trust offices 231, 232, **233**
rooftop sensors 141
RT *see* reverberation time
run-around coil 132, **133**, 134

SAD *see* seasonal affective disorder (SAD)
St John's College, Cambridge 44, **131**
salt water bath modelling 169, **172**, 196
Salvation Army Hotel, Paris 43
SBS *see* sick building syndrome
schools
   background noise 20
   energy consumption *32*
   hot water service demand 146–7, *146*
   overheating 31, 38
   temperature 14
   Victorian 73
   water expenditure 145
screw anchors 213, **213**
sealants 45
seasonal affective disorder (SAD) 105
Seasonal Efficiency of Domestic Boilers in the UK (SEDBUK) 112
seasonal heat store 226–9, *226*
SEDBUK *see* Seasonal Efficiency of Domestic Boilers in the UK
selection of materials 67
semiconductors 90
sensible heat 7, 111
   light level 174
   temperature 138–9, **138**, 220
   ventilation 141
sensors, building management systems 178
septic tanks 150
service sector, carbon dioxide emissions 29, *30*
services materials 71–2, 76
shading 43–5, **43**, **44**, 158, 160, 201, **233**
   PV panels 231, **233**
shelterbelts 62–3, **63**, 64
sick building syndrome (SBS) 23
single-sided ventilation 49, **121**, 124, 125
sinusoidal slab 194–6, **194**
site planning 54–66
skylights 100–2, **100**, **101**, **102**, 158, 160
   National Trust offices 231, 232, **233**

smoke detectors 141
smoke ventilation 129
smokeless fuels 84
smoking 23
solar access 57, **57**, **59**, 63
solar energy 12, 179, 191, 192
   active solar heating 88–9
   Beaufort Court 224–9, **224**, **225**, *226*, **228**
   photovoltaic and solar thermal panel (PVT) 225–6, *226*
   photovoltaics 88, 90, 94, 255, 255–60, **255**, *258*, **259**
   BRE Environmental Building 203–5, **203**, **204**, 208
   Millennium Centre 214
solar gain 31, 33–4, 43, 56–60, *168*, 200–2, 214, 231
solar radiation 3–4, 6, 15–16, 57, 88–90
   see also daylight; solar energy
   absorptance 12
   altitudes/azimuths *243*
   annual mean **242**
   energy flows 31, 33–4
   glass characteristics 41, *42*
   incident **259**
   sunshine hours *241*, **241**
   windows 40–5
solar shading 223
'solar stacks' 196
solar water heating 88–9, **88**, **89**
sound
   see also acoustics; noise
   comfort factors 18–21
   insulation 20–1, 21, **21**, *42*, 176
   representative sound pressure levels *253*
space heating *see* heating
specific heat capacity 9, *11*
spectral distribution
   artificial lighting **103**
   radiation 15, **15**
spectral responsivity 245, **255**
spectral transmission curves **41**
specular reflection 18
stack effect ventilation **10**, 119–26, **121**, **126**, 129
   Bedales Theatre 218–20
   BRE Environmental Building 195, **195**
   De Montfort University Queens Building 125, 168–9, *170–1*, 172–3, *172*, **173**
   Millennium Centre **212**, 214
   National Trust office **233**, 234–5
standard overcast sky illuminance 97

stiffness 73, *74*
storage of hot water 148–9
stratification
    temperature 161
    water circulation systems 149
straw, as fuel 93, *93*
structural materials 73–4
sulphur dioxide (SOx) 22, 23, 84,
    113–14
sunlight *see* daylight; solar gain; solar
    radiation
sunshine hours *241*, **241**
surface water run-off 56
sustainable development 68, 211
    National Trust brief 231, 234, 235–6
Sustainable Urban Drainage 56
swales 56

taps, water conservation 145
technology 3, 138
    see also automatic control systems;
        computers
temperature
    see also heat recovery; heating
    admittances 37–40
    building management systems 200
    comfort 7–14, 192
    considerations **110**
    De Montfort University Queens
        Building 177, 179
    dry resultant 234–5
    electromagnetic spectrum 11, 15–17
    hot water storage 149
    internal summertime **207**
    microclimate 55
    'occupant' 234–5
    photovoltaics 259
    psychrometric chart **247**
    RMC House 161, 162, **162**
    seasonal figures *243*
    sensors 220
    units xv
    ventilation 4, 123–6, *126*, *127*
TFS *see* thin film silicon
thermal balance 13
thermal comfort 8–14, 110–44, 192
thermal conductivity 9, *11*, 37, *75*, 250
    see also conduction
thermal mass 9, *11*, 37, 39, 51, 231
    Bedales Theatre 218
    BRE Environment Building 193
    Millennium Centre 213
    RMC House 158, 163
thermal resistance *75*, *166*, 249–50
thermal response 114, **116**
thermal wheels 132, **133**

thermochromic glass 42
thermostatic radiator valve (TRV) **139**
thin film silicon (TFS) panels 203
timber 68, 71, 73–4, 77, 193, 206
Timber Research and Development
    Association (TRADA) 74
town gas 83
toxins 70
TRADA *see* Timber Research and
    Development Association
trees 59–60, **59**, *60*, 68, 223
trickle ventilators 45, *46*, 135, 197
triple-glazing 6, 177, 213
TRV *see* thermostatic radiator valve
tungsten lamps 76, 104, *106*
turbines, wind energy 55

U-values
    Bedales Theatre 218
    BRE Environmental Building 193
    De Montfort University Queens
        Building *166*
    glazing 41–2, *42*, 45, 49
    Millennium Centre 212
    National Trust central office, Heelis
        249–50
    RMC House 158
    wind resistance 62
UE *see* useful energy
ultraviolet radiation 15–16
underfloor heating coils 197
units xv–xvi
University of East Anglia 136
University of Northumbria,
    Northumberland Building 257, 258,
    260
uplighting 105, 107, 161
useful energy (UE) 81, *82*

vacuum cleaning 24
vapour compression air-conditioning
    138
vector/scalar ratio 104
vegetation 59–60, **59**, *60*, 62, 64
velocity pressure (VP) 122
ventilation 4, 45, 118–38
    see also cooling systems; infiltration
    air flows **121**, **122**
    air quality 23
    assisted natural 132–5, 140–1
    Bedales Theatre 218–20, **219**
    BRE Environmental Building 193–6,
        194, **195**, 197
    atria 49
    attenuation 46, *47*
    condensation 77

considerations **110**
controls 140–1, **140**, 177–9, 234
    with cooling 137–8
    costs *181*
    cross 119, **121**, 123–4, 125
    atria 49
    BRE Environmental Building 195
    De Montfort University Queens
        Building 169, *170*
    Millennium Centre **212**, 214
    energy flows 33
    heat loss 235
    mechanical 49, 135–7, 141, 193
    Charles Cryer Studio Theatre
        186–8, *186*
    RMC House 158, 160–1, **160**,
        163
    mechanical strategy **236**
    microclimate 55
    natural 49, 51, 55, 62, 119–35
    checklist *127*
    controls 140–1
    costs *181*
    De Montfort University Queens
        Building 168–73, *170–1*, *172*, **172**,
        173
    National Trust office 231, **233**,
        234–5
    night *43*, *44*, 193
    performance assessment 235
    'snouts' **233**, 234
    stack effect **10**, 119–26, **121**, *126*,
        129
    Bedales Theatre 218–20
    BRE Environmental Building 195,
        195
    De Montfort University Queens
        Building 125, 168–9, *170*, *172*,
        172, **173**
    Millennium Centre **212**, 214
    National Trust office **233**, 234–5
    thermal mass 39
    wind 62, 123, 168–9, *170–1*
ventilation, night 231, 234
vernacular architecture 119
Versailles Palace 70
views 40, 61
vision, special response **17**
voids, light distribution 231, 232, **232**

walls 21, 74, 77, 212
warm air heating systems 114–15, **116**
waste
    agricultural 93, *93*
    demolition 223
    disposal 150

municipal, renewable energy 94
recycling 216
reduction 78
water
see also hot water service
atmospheric water vapour pressure 7
condensation 25
conservation 146
consumption reduction 236
dilution 24–5
evaporative heat transfer 13
groundwater cooling systems 138
BRE Environmental Building 191, 198–200, **198**
Charles Cryer Studio Theatre 184–8, *184*, *185*, *186*, *187*
hydroelectric power 92

microclimates 55
recycling 216
surface run-off 56
Water Supply (Water Fittings) Regulation 1999 145
WCs 145, 150
wind 4, 54, 61–4
annual distribution **191**
Beaufort scale 244, *245*
isopleths of wind speeds **246**
renewable energy 88, 90–2, *91*, **92**, 94, 214–15, **214**, **215**
roses **246**
turbine 224–5, *226*
ventilation 123, 168–9, *170–1*
wind energy 55
windbreaks 62–3, **63**, 64
windmills *91*, **92**

windows
see also glazing
composite 47–9, **48**
daylight 97
energy exchange 40–2, **41**
heat loss 12
responsive glass 40
solar radiation 6, 40–5
ventilation 129, **130**, 234
wood
see also timber
as fuel 92
work 81
workplace health issues 70

zoning
lighting 105, 107
thermal controls 138, 228